FANTASTIC VOYAGES

VOYAGES

Learning Science Through
Science Fiction Films

FANTASTIC VOYAGES

Learning Science Through Science Fiction Films

Leroy W. Dubeck
Suzanne E. Moshier
Judith E. Boss

LIBRARY
DEXTER MUNICIPAL SCHOOLS
DEXTER NEW ...

Springer

Cover Illustration: Artist's concept of an intermediate size (300-meter) space SETI (Search for Extraterrestrial Intelligence) system antenna showing two feeds, a relay satellite, RFI (radio frequence interference) shield and a shuttle type vehicle. Located in geosynchronous orbit or beyond. Courtesy of NASA.

©1994 by the American Institute of Physics.
All rights reserved.
Printed in the United States of America.
Second Printing 1995.
Third Printing 1995.

Reproduction or translation of any part of this work beyond that permitted by Section 107 or 108 of the 1976 United States Copyright act without the permission of the copyright owner is unlawful. Requests for permission or further information should be addressed to the Office of Rights and Permissions, 500 Sunnyside Blvd., Woodbury, NY 11797-2999; phone: 516-576-2268; fax: 516-576-2499; e-mail: rights@aip.org

In recognition of the importance of preserving what has been written, it is a policy of the American Institute of Physics to have books of enduring value published in the United States printed on acid-free paper, and we exert our best efforts to that end.

AIP Press
American Institute of Physics
500 Sunnyside Boulevard
Woodbury, NY 11797-2999

Library of Congress Cataloging-in-Publication Data

Dubeck, Leroy W., 1939–
    Fantastic Voyages: Learning science through science fiction films/
Leroy W. Dubeck, Suzanne E. Moshier, and Judith E. Boss.
    p.   cm.
Includes bibliographical references.
ISBN 1-56396-195-4
    1. Physics—Study and teaching—Audio-visual aids. 2. Biology—Study and teaching—Audio-visual aids. 3. Science fiction films. I. Moshier, Suzanne E. II. Boss, Judith E. III. Title.
QC30. D83 1994   93-10039
500—dc20   CIP

Printed and bound by Electronic Printing Inc., Plainview, NY.
Printed in the United States of America.

9 8 7 6 5 4

ISBN 1-56396-195-4   SPIN 10830863

Springer-Verlag New York Berlin Heidelberg
A member of BertelsmannSpringer Science+Business Media GmbH

TABLE OF CONTENTS

PHYSICS

BIOLOGY

FILM DESCRIPTIONS

ACKNOWLEDGMENTS

The authors wish to thank Rose Tatlow and Jennifer Steinberg for writing a number of the film descriptions and for editorial assistance in preparing the final draft of the manuscript. The authors also are indebted to Marjorie Mendelson for manuscript typing and editing. Finally, the authors are grateful to Robert Singer for drawing all of the illustrations.

The still photographs were provided by the Museum of Modern Art/Film Still Archives, courtesy of Allied Artists Pictures, Ladd Company, Metro–Goldwyn–Mayer, RKO, 20th Century Fox, Universal Pictures, Universal International, and Warner Brothers.

The authors wish to thank Teachers College Press for permission to reprint the film descriptions of *Forbidden Planet, The Day the Earth Caught Fire, The Day the Earth Stood Still, The Day of the Triffids, Them!, The Andromeda Strain*, and *Colossus: The Forbin Project*. These film descriptions first appeared in *Science in Cinema: Teaching Science Fact Through Science Fiction Films* by Leroy W. Dubeck, Suzanne E. Moshier, and Judith E. Boss, published by Teachers College Press of Columbia University, New York, copyright 1988. All rights reserved.

This material is based upon work supported by the National Science Foundation under Grant No. USE-9055563. Any opinions, findings, conclusions, or recommendations expressed in this publication are those of the authors and do not necessarily reflect the views of the National Science Foundation.

INTRODUCTION

Do you relish science fiction films but do not have quite the same feeling about science? If so, this book was written with you in mind. We, the authors, are college professors who also happen to love science fiction films. Because of our scientific expertise, we can get even more enjoyment from screening science fiction films than the average viewer because we understand what is possible or what is truly "out of this world" in the universes portrayed on film.

After reading this book and working through the exercises at the end of each chapter, you will get more out of screening science fiction films—you will have a deeper understanding of the scientific principles presented in the films and probably a new respect for those filmmakers who depict science fiction that is close to science fact.

The goal of this book is to provide basic physics and biology instruction, using scenes from science fiction films as examples of the concepts discussed. Your instructor will most likely use clips from the films to illustrate a scientific principle, and you will probably find that seeing a portrayal of a scientific principle (or of a principle's violation) helps you to understand a concept better than more traditional methods of classroom instruction.

Furthermore, by watching films you will get a better feel for how different fields of science interact—rarely does a film deal with physics only and not for example, biology, astronomy, or the social sciences.

We hope that you will come to enjoy both science and science fiction films as much as we do. (Even if you have to forego the popcorn!)

How to Use This Textbook

In the Table of Contents we list the page numbers of films discussed in the textbook. Films are also listed in the index.

This book is divided into three sections. The first section discusses basic physics and astronomy for nonscience majors. The second is devoted to selected topics in biology. The third section of this book gives detailed plot descriptions of 32 films which are referenced in the physics and/or biology sections. The first 15 of these films also have sections describing relevant science fiction literature, which may be used in conjunction with a given film and the related physics, astronomy, or biology material.

At each point in the physics/astronomy section at which a film is introduced to illustrate a concept or its violation, the running time is given at which the referenced segment of that film begins. For example, in Chapter One, *The Andromeda Strain* is referenced. The introduction to the film segment describing the attempt to determine the size of the alien organism is followed by (69 min). This means that the film segment starts 69 min after the film commences. One can precisely locate the starting point of a given film segment by using a VCR that indicates the running time of the film at any point on the tape. Alternatively, one can estimate the position on the tape at which the segment commences by taking the ratio of the running time at which the segment commences to the total running time of the film. For example, if the segment commences at 60 min and the running time of the entire film is 120 min, one needs to fast forward the tape until 50% of it has elapsed.

In the biology section, the films utilized normally relate to a number of biological topics. These films are therefore introduced at the conclusion of each chapter rather than appearing throughout the chapter. Running times are usually not given for film segments since several segments of the film are often related to the material in the chapter.

A number of exercises and discussion questions are presented at the end of each chapter.

Many of the films are used in both the physics/astronomy section and the biology section. These films demonstrate the inherent interdisciplinary nature of science.

There is an instructor's manual which contains answers to all the exercises available for this book. It can be ordered free of charge from AIP Press, Maria Taylor, 500 Sunnyside Blvd., Woodbury, NY 11797.

PHYSICS

Chapter One
Science

A Brief History of Science

The goal of the sciences is the ordering of the universe detected by our senses. The nonscientist may think of the sciences as simply processes by which to collect facts and formulate theories. But science is more than that: it is a creative activity that, in many respects, resembles other creative, yet ordered, activities such as art, literature, and music.

The body of knowledge that is incorporated under the heading of science had its beginnings before recorded history, when regularities in nature were first observed, such as when the rainy season started. From these regularities our ancestors began to make predictions that gave them some sense of control over nature.

The field of science grew greatly during the Grecian expansion of more than 2,500 years ago and spread throughout the Mediterranean world. European scientific advances came to an abrupt halt, however, when the Roman empire collapsed in the fifth century A.D. Barbarian hordes ravaged Europe and ushered in the Dark Ages. Fortunately, science and mathematics flourished in China and the Arab nations during this period. Science, as practiced by the ancient Greeks, was reintroduced to Europe by Islamic peoples starting in the tenth century A.D. European universities emerged in the thirteenth century. The introduction of gun powder in the fourteenth century changed warfare, and thus provided a concrete example of the practical importance of scientific investigations. The advent of the printing press in the sixteenth century made scientific discoveries more widely known. During that century, the Polish astronomer Copernicus (1473–1543) caused great controversy when he published a book proposing that the Sun was stationary and that the Earth revolved around it. This was in contradiction to Church teaching and Copernicus, fearful of the Church's reaction to his theory, published his book only in the final days of his life. A century later, the Copernican theory of the solar system was accepted.

It is interesting to note that Copernicus' Sun-centered theory of the solar system was no more accurate than the earlier Earth-centered theory of Ptolemy (second Century A.D.) for predicting the motion of heavenly bodies. Copernicus' theory had wider application, however, because it explained a greater range of phenomena than Ptolemy's theory. For example, Copernicus' theory made possible a determination of the order and distance of the planets. It was also a simpler theory than that of Ptolemy, who assumed that heavenly bodies moved in circles that, in turn, moved around yet other circles. Wider applicability, simplicity, as well as quantitative agreement, play a major role in the acceptance of a scientific theory.

Science, Religion, and Technology

The human desire to explain the world around us has taken us on many different paths: one path is science and another is religion. Although both seek to explain their domains, they are different. Science is mainly engaged with discovering and understanding natural phenomena, while religion addresses the source, purpose, and meaning of life.

Science is often confused with technology, but these fields, too, are very different. Science is concerned with discovering the relationships between observable phenomena and with organizing and describing these phenomena in terms of theories. Technology is concerned with the tools, techniques, and procedures for putting to use the discoveries of science.

Another difference between science and technology is their impact on human beings. Unlike science, advances in technology must be measured in terms of their impact on humanity. In addition, one has the option of refusing to believe a scientific theory such as the planetary model of the atom. This is hardly the case once a theory is applied to technology: one does not have the option of living in an age without nuclear power plants and nuclear weapons.

Some people, perhaps frustrated with the shortcomings of science, have turned to pseudoscience for answers. Pseudoscience may be defined as a belief in physical phenomena which is not supported by the scientific method. As an example, some people believe that there are individuals who can move things with their minds. As we shall see later, there is no convincing evidence that this ability exists, and overwhelming arguments that it does not. Yet the belief in this phenomenon continues on the part of millions of people.

The Scientific Method

The Italian physicist, Galileo Galilei (1564–1642) was one of the principal founders of the **scientific method**. This method can be described as follows:

(1) recognize that a scientific problem exists.
(2) state a hypothesis, namely, make an educated guess about the problem.
(3) predict the consequences of the hypothesis.
(4) perform experiments to see if the predictions occur.
(5) formulate the simplest general rule that organizes the hypothesis, predictions, and experimental results into a **theory.**

A scientific theory is a synthesis of well-tested and verified hypotheses about some aspects of the world around us. For example, physicists work with the theory of the atom, and biologists work with the theory of cells.

When a scientific hypothesis has been tested over and over again and has never been contradicted by experimental results, it may become known as a **scientific law or scientific principle**.

A **scientific fact** may be defined as an agreement by competent observers of a series of observations of the same phenomena. Sometimes facts are revised by additional data about the world around us. For example, it is now a scientific fact that the Earth is round. It was once considered a fact that the Earth was flat.

Scientists often employ a **model** in trying to understand a particular set of phenomena. A model is an analogy or mental image of the phenomena in terms with which we are familiar. For example, there is the "planetary model" of the atom in which scientists visualize the atom as a nucleus with electrons revolving about it, just as the planets revolve around the Sun. While this model is useful in understanding the atom, it is important to understand that it is an oversimplified description of an atom and thus does not predict some of its attributes.

Science and the Media

An extensive source of both science and pseudoscience for the public is the media. Throughout this book we shall employ science fiction films and science fiction television programs to illustrate scientific principles or their violation. For example, the film, *The Andromeda Strain*, describes the efforts of a team of scientists to deal with an extraterrestrial organism which is fatal to most humans. The organism apparently had been brought to Earth on a satellite which had crashed in a small town. All but two of the inhabitants died soon after the capsule was opened. The satellite's contents are carried to a top secret research installation at which scientists are faced with the problem of determining what the organism is and how to control it. Refer to the plot description for more details.

The scientific method is well illustrated by the attempt in the film to determine the properties of the organism (59 min). The scientists form a hypothesis that the organism is airborne, which would explain why it quickly killed nearly everyone in town. They test this hypothesis by exposing a test animal to air that is in contact with the capsule. The animal's immediate death confirms the hypothesis that the organism is transmitted via the air.

The scientists then try to determine the organism's size (69 min). Is it a gas, a virus, or something larger? The scientists put a series of filters between a sealed case containing the first test animal which had been exposed to the capsule and another test animal and gradually increase the size of the holes in the filter until the second animal dies. They thus determine that the size of the organism is between 1 and 2 μm in diameter since it passes through 2-μm-diam holes but not through 1-μm-diam holes. (1 μm is one-millionth of a meter.) Therefore, it is large enough to be a cell.

Measurement and Uncertainty

Science is quantitative because it involves the measurements of physical properties. Every measurement has a level of uncertainty associated with it. For example, if you used a centimeter (cm) ruler to measure the width of a book, the result might be accurate to about 0.1 cm, the smallest division on the ruler. The reason for this claim of accuracy would be the difficulty in interpolating between the smallest divisions on the ruler. Whenever one gives the result of a measurement, it is good practice to state the precision, or **estimated uncertainty**, in the measurement. For example, the width of the book would be written as 18.1 ± 0.1 cm. The ± 0.1 represents the estimated uncertainty in the measurement so that the actual width of the book lies between 18.0 and 18.2 cm.

The number of reliably known digits in a number is called the number of **significant figures**. There are three significant figures in the number 18.1 cm, but only two in the number 0.018 cm, since the zeros in the latter are merely place holders. When making measurements or performing calculations, one should not keep more significant figures in the final answer than the number of significant figures in the least accurate factor in the calculation. For example, the calculation of the area of a rectangle which is 1.05 cm by 6.6 cm would yield 6.93 cm^2. But the answer is accurate to only two significant figures since one of the factors (6.6) is known to only two significant figures. Thus the correct answer is 6.9 cm^2.

Scientific Notation

Since the result of the measurements (or estimates) of certain physical quantities are either very large or very small numbers, scientists need a compact way of writing these numbers. These numbers are compressed by using **powers of ten** rather than a string of zeros following the number. For example, $2,000,000,000 = 2 \times 10^9$. The power of ten (i.e., the raised number to the upper right of ten) gives you the number of zeros following the number.

Therefore $3 \times 10^6 = 3,000,000$ since the number 6 means there are six zeros after the 3. Another way of looking at this is to consider 10^6 as $10 \times 10 \times 10 \times 10 \times 10 \times 10$, or 10 multiplied by itself 6 times. Large numbers having more than one digit (excluding the zeros) are treated as follows.

(1) Start with the number as given. Place the decimal point after the first digit. Then count the number of places that the decimal point has been moved to the *left*. This gives you the exponent of ten.

Examples:

$$314,000 = 3.14 \times 10^5,$$

$$2,895,000,000 = 2.895 \times 10^9.$$

Note that one continues to write all of the nonzero numbers. You change only the position of the decimal point and replace zeros by a power of ten.

This is the way scientists express large numbers such as the number of molecules in a room or the number of stars in a galaxy. What about very small numbers such as the size of a single molecule? These numbers can also be expressed as a power of ten. Here the decimal point moves to the *right* and is represented as a negative exponent of 10.

Example: $0.00000543 = 5.43 \times 10^{-6}$, since the decimal point was moved six places to the right. Once again, you continue to write all of the nonzero numbers. Another way of looking at negative numbers for exponents of 10 is to realize that

$$10^{-6} = \frac{1}{10^6} = \frac{1}{1,000,000} \, ,$$

or one-millionth. Note that the number one is always implied; in other words, $10^6 = 1 \times 10^6 = 1,000,000$. This notation will be used throughout this book.

Go through the following additional examples:

$$10^1 = 10$$

$$10^2 = 100$$

$$10^3 = 1,000$$

$$10^4 = 10,000$$

$$10^5 = 100,000$$

$$2.42 \times 10^8 = 242,000,000$$

$$3.57 \times 10^5 = 357,000$$

$$2 \times 10^3 = 2,000$$

$$4.57 \times 10^{-6} = 0.00000457$$

$$3.14 \times 10^{-3} = 0.00314$$

Units and Standards

The measurement of any physical quantity is made in terms of a particular standard or unit, and this unit must be specified along with the numerical value of the quantity. For example, length is measured in units such as inches, feet, miles, or meters. To specify the width of a book, you cannot write just 18.1. Rather, you must write 18.1 cm.

Several systems of units have been in use over the years. Today, the most important is the Systeme International (French for International System) (SI). In SI the standard base unit for length is the **meter**, the standard base unit for time is the **second**, and the standard base unit for mass is the **kilogram**. There are seven base units in the SI system: we will discuss some of the other four base units later in the book. Most readers will be more familiar with the British engineering system of units, which takes as its standards the foot for length, the pound for force, and the second for time. This book will use units from both systems.

The SI standard unit of length, the meter, was originally chosen to be one ten-millionth of the distance from the Earth's equator to either pole. In 1960, the meter was redefined to be a certain number of wavelengths of a particular orange light emitted by the gas krypton 86. In 1983, the meter was again redefined in terms of the distance traveled by light in a vacuum during a specified period of time. The inch is currently defined as precisely equal to 0.0254 m (or 2.54 cm).

The SI standard unit of time is the second. It was originally defined as 1/86,400 of a mean (average) solar day. The standard second is now more precisely defined as a multiple of the period of a specified radiation.

Finally, the unit of mass in SI is the kilogram (1 kg weighs about 2.2 lb on the surface of the Earth).

The physical quantities length, time, and mass are considered **base quantities**, since they must be defined in terms of a standard. By contrast, other quantities can be defined in terms of these base quantities, and hence are referred to as **derived quantities**.

One must be careful to always use a consistent set of units in doing a calculation. For example, in determining the distance that one travels in a given period of time, if speed (which is equal to distance traveled divided by time, and is an example of a derived quantity) was stated as being 60 miles per hour (mph) and you are asked to determine how far one would travel at this speed in 3 h, you would have to multiply 60 mph×3 h=180 miles to get the correct answer. However, 3 h is equal to 180 min. If you were to multiply 60 mph×180 you would get a ridiculous answer for the distance traveled because you would have used an inconsistent set of units, namely, both hours and minutes for time in the same equation.

Changing Units

It is sometimes necessary to change from one set of units to another. Units can be changed by applying a simple trick: multiplying the given quantity by one. For example, 1 h is equal to 3,600 s. Thus, a quantity multiplied by (1 h)/(3,600 s), is multiplied by one.

Let us apply this technique to convert a speed of 60 mph to ft/s.

60 mph=(60 mph) (5,280 ft/mile) (1 h/3,600 s)=88 ft/s.

When changing units it may be unclear whether to use the conversion factor (1 h)/(3,600 s) or the conversion factor (3,600 s)/(1 h). Writing out the conversion factor will help one to see which version will permit the cancellation of the unwanted units, such as miles and hours in the above example.

Exercises

1. Should a scientist ever change a theory once it has been stated and accepted by the scientific community? Explain.

2. In daily life, people who are caught misrepresenting things are sometimes thereafter excused and accepted by their contemporaries. Is this different in science?

3. How would you describe the following statement: "A large asteroid struck the Earth many millions of years ago, causing debris to be hurled into the atmosphere which cut off sunlight, caused the deaths of plants across the globe, and therefore led to the extinction of the dinosaurs." Is this a hypothesis, a theory, or a physical law? What kind of supporting evidence for it would you seek?

4. State the following numbers in scientific notation:

186,000,

92,000,000,

0.0000357.

5. How can a "fact" of science change?

Chapter Two
Mechanics

Mechanics deals with how and why objects move. Mechanics is usually divided into two parts, **kinematics**, which is the description of *how* objects move, and **dynamics**, which deals with *why* objects move. We start first with kinematics.

Speed and Velocity

Speed and velocity are words often used interchangeably in everyday life. In physics, they have different meanings. Speed refers to how far an object travels in a given time interval. The **average speed** of an object is defined as follows:

average speed=distance traveled divided by time elapsed.

This can be rewritten in terms of symbols as

$$\bar{v} = d/t,$$

where v is the average speed, d the distance traveled, and t the time elapsed. The bar over the v is the standard symbol for "average." As an example, if you drove 200 miles in 4 h, your average speed=(200 miles)/(4 h)=50 mph.

The speed that an object has at any given instant of time is called the **instantaneous speed**.

The total distance covered equals the average speed times the time elapsed. This comes from a simple rearrangement of the definition of average speed. As an example, if your average speed is 100 km/h on a 5 h trip, you will cover a total distance of 500 km.

Speed simply states how fast an object is going. **Velocity, v**, is speed in a given direction. The boldfaced symbol indicates that the quantity is a **vector**, something which has both a magnitude as well as a direction. The magnitude of a vector is its size or measurable quantity. For example, the magnitude of velocity is speed. Velocity gives us more information about an object's motion than simply stating its speed. Thus, stating that a car is going 60 mph, i.e., stating its speed alone, tells nothing about the direction in which it is moving. On the other hand, stating that the car is moving 60 mph north determines its velocity. Physically, the direction of motion is as important as its magnitude, as we shall later see.

One can define average velocity, as well as instantaneous velocity, by analogy to speed. However, constant velocity and constant speed can be very different. Constant velocity means constant speed with no change in direction. On the other hand, a car that rounds a curve at constant speed does not have a constant velocity because its direction of motion constantly changes.

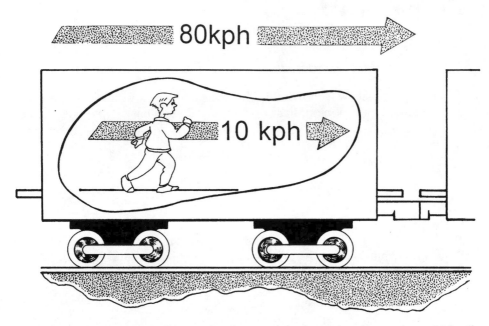

FIGURE 2-1. A passenger walks at 10 km/h toward the front of a train moving at 80 km/h.

Frame of Reference

Speed, distance, etc. must all be measured with respect to a **frame of reference**. For example, a passenger walking at 10 km/h toward the front of a train moving at 80 km/h down railroad tracks has a speed of 10 km/h with respect to the train as a coordinate system but also has a speed of 90 km/h with respect to the railroad tracks, as pictured in Fig. 2-1. Thus, stating a person's speed is meaningless without giving the frame of reference with respect to which the person's speed is measured.

In physics, we often draw a coordinate axis, as shown in Fig. 2-2, to represent a frame of reference. The point at which the coordinate axes intersect is referred to as the origin (o) of the coordinate axes. Positions to the right of the origin have an x coordinate with a positive value, while points to the left of the origin have a negative x coordinate. The positions along the y axis are usually considered positive when above the origin, and negative when below it. Any point on the xy plane can be uniquely specified by giving its x and y coordinates. Finally, although most measurements are made in reference frames fixed on the Earth, other reference frames are perfectly legitimate, such as reference frames fixed on moving ships or on other planetary bodies.

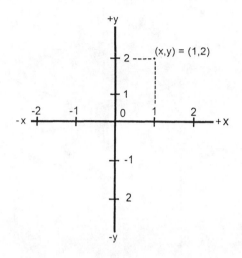

FIGURE 2-2. A coordinate system which locates a point in a plane by giving its *x* coordinate and *y* coordinate.

Acceleration

We can change the velocity of an object by changing its speed, by changing its direction, or by changing both its speed and direction. **Acceleration** is defined as the rate of change of velocity.

Acceleration — change of velocity/time interval.

As an example, let us consider that a man increases the speed of his car driving down a straight road from 60 to 80 ft/s in 4 s. Then, $a = (80 - 60)$ ft/s/4S = 5 ft/s/s, which can be written more simply as 5 ft/s^2. The direction of the acceleration in this example is the direction of velocity.

Acceleration can also occur when the direction of velocity changes even though the magnitude of the velocity, i.e., the speed, remains the same. As an example, imagine riding on a merry-go-round, rotating at constant speed. You would feel yourself being "pulled" off the merry-go-round, and would have to hold on to avoid falling off. This pull occurs because you are experiencing an acceleration as the merry-go-round turns and, as we shall see shortly from Newton's First Law, your body wants to continue to move at constant speed in a straight line. To do this would require you to fly off the merry-go-round. In order to avoid falling off, you have to hold on to the merry-go-round so that it can pull you in a circle. Your speed is the same but the direction of your speed constantly changes, as indicated in Fig. 2-3.

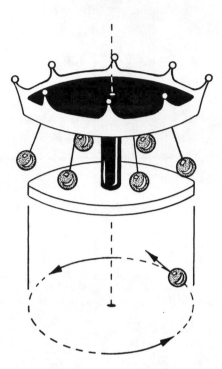

FIGURE 2-3. Riders on a merry-go-round are continuously accelerated. Their speed is the same everywhere but the direction of their speed continuously changes.

Newton's Laws of Motion: Dynamics or Why Objects Move

A concept which is important in describing the way objects move is that of **mass**, which is defined to be the quantity of matter in a body. Mass is the quantitative measure of the inertia, or sluggishness, that a body exhibits in response to any effort to change its velocity. Mass is distinguished from the weight of the body, which is the force of gravity acting upon the body due to the planet on which the body is located. Thus, the weight of an astronaut on the Moon is only one-sixth his weight on Earth, but the astronaut's mass is the same on both planetary bodies, since the quantity of matter in the astronaut remains unchanged in moving from the Earth to the Moon.

By the time Isaac Newton (1642–1726) was 23 he had developed his famous laws of motion that overthrew the ideas of the Greek philosopher, Aristotle (384–322 B.C.), which had dominated Western science for over 2,000 years.

Aristotle had concluded that a moving object must have a force exerted on it to keep it in motion. He based this on the observable fact that when you take an object, e.g., a book, and push it, it will soon come to rest unless you continue to exert a force (a push or a pull) upon it. Therefore, Aristotle concluded that the natural state of things was to be at rest. This also led him to conclude that the Earth must be at rest and at the center of the Universe; other heavenly bodies, including the Sun, revolved around the Earth.

Aristotle's error in reasoning was caused by his overlooking the role that friction plays in determining the motion of objects. The book moving across the table comes to rest because of forces exerted on the bottom of the book by the top of the table. These forces cause the book to slow down and eventually stop. If one could reduce the force of friction by pushing a very light object across the top of an air table where it rides on little jets of air (which allows the object to move without physically touching the table itself), one would find that the object slows down much less quickly than the book on the table. Eventually, an object will come to rest even on an air table because there is some friction between the object and the air molecules.

Newton reasoned that an object would continue moving in a straight line at a constant speed forever, if it was moving across a surface free from all friction.

Newton's First Law of Motion states:

Every body continues to remain at rest, or to move at constant speed in a straight line, unless acted upon by a resultant force.

A "resultant" or "unbalanced" force is one which is not counteracted by another force. For example, if you push forward on a carton on the floor, it will move once the force you apply is larger than the force of friction between the bottom of the carton and the floor. That frictional force opposes any motion of the carton and is in the direction opposite to any applied force which attempts to move the carton. As long as the frictional force is equal in magnitude (but opposite in direction) to the applied force, there is no resultant force on the carton and it does not move. Similarly, a rocket ship in interstellar space, far away from the gravitational pull of any planet or star, would continue to move in a straight line at a constant speed, even with its rockets turned off.

Newton next considered the case where there was a resultant force, **F**, acting on an object of mass, m. **Newton's Second Law of Motion** is as follows:

$F = ma$, where **F** is the resultant force acting on the object and **a** is the acceleration produced by the force. Note that both **F** and **a** are vectors. They each have a direction as well as a magnitude. This law is consistent with everyday experience. If you push forward on a box, it moves forward. In other words, it is accelerated from rest in the forward direction. It does not accelerate backward against the applied force.

In the SI system of units, kilograms are used for mass, meters/seconds^2 for acceleration, and newtons (N) for force. The source of the resultant force may be gravitational, electrical, magnetic, or simply a muscular effort pushing or pulling an object. We say that **F** represents the resultant force because there are often many forces acting on an object. Consider this concept in detail. If two equally strong individuals were to pull an object in the same direction, the resultant force would be twice as great as the individual force exerted by each person. If the two individuals were to pull the object in opposite directions, the forces would cancel;

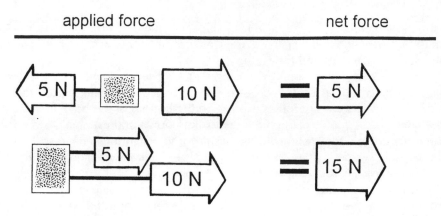

applied force   net force

FIGURE 2-4. The net force acting on an object if a 5 N and a 10 N force act either in opposite directions or in the same direction.

thus there would be zero resultant force acting on the object, and from Newton's Second Law, there would be zero acceleration on the object. Thus, if the object were initially at rest, it would remain at rest because a zero acceleration means that the velocity of the object does not change.

Suppose, on the other hand, that a force of 10 N was exerted on an object and a second force of 5 N was also exerted on the same object. Then, the net force would be 15 N if both forces pulled in the same direction, but only 5 N if the forces pulled in opposite directions, as illustrated in Fig. 2-4.

One can solve for any one of the quantities in Newton's Second Law, if one is given values for the other two symbols. For example, if one applies a force of 20 N to a ball of mass 0.4 kg, the ball will receive an acceleration, in the direction of applied force, of

$$\mathbf{a} = \mathbf{F}/m = 20 \ N/0.4 \ kg = 50 \ m/s^2.$$

One should note that Newton's First Law can be considered as a special case of Newton's Second Law. When the resultant force on an object is zero, its acceleration is zero, and hence its velocity will be unchanged, i.e., it will continue to remain at rest or to move at constant speed in a straight line.

Newton's Third Law of Motion states:

For every force there is an equal and opposite force. Or, whenever one object exerts a force on the second, the second object exerts an equal force in the opposite direction on the first object.

Thus, when you push against a carton, you can feel the carton pushing back against your hand. If you kick a rock, the rock exerts an opposite force on your shoe.

Newton's Third Law deals with the origin of forces in contrast to Newton's First and Second Laws which deal with the effects of forces. The Third Law always

FIGURE 2-5. The forces exerted on a horse-drawn cart: F_e is the force of the Earth on the horse, F_h is the force of the cart on the horse (which is equal to magnitude to the force of the horse on the cart), and F_c is the frictional force on the wheels of the cart.

applies to two objects and states that forces exist in pairs. In other words, you cannot exert a force on an object without the object exerting a force back on you. When you walk across the floor you push back against it, and the floor pushes forward against you. Both you and the floor push against each other simultaneously. In swimming, you push against the water backward, which, in turn, pushes you forward. This principle is also basic to the motion of all rockets. The rocket pushes on gas molecules which it ejects. In turn, the gas molecules push on the rocket which makes the rocket go forward. Note that the two forces *never* apply to the same object.

This law can be confusing. For example, (see Fig. 2-5) if a horse pulls forward on a cart, the cart pulls backward with an equal but opposite force on the horse. How can the cart ever move? It may appear that the horse's force is cancelled by the reverse force of the cart, leaving a total of zero force. What this argument overlooks is the fact that the two forces act on *different* objects. The cart has only one force acting on it due to the horse, F_h. Therefore, it is accelerated in a forward direction if F_h is larger than the frictional force, F_c, acting in the backward direction on the wheels of the cart. As for the horse, there are actually two forces acting on the horse in the horizontal direction. There is the backward force, F_h, due to the cart and a forward force, F_e, due to the Earth pushing forward on its hoofs. Thus, as Fig. 2-5 indicates, when the horse presses backward on the Earth with its hoofs in order to start pulling the cart, by Newton's Third Law the Earth pushes forward on its hoofs. Once this F_e exceeds F_h the horse is accelerated in the forward direction by the Earth! Since F_h exceeds F_c, the cart also is accelerated in the forward direction.

While the forces are the same in magnitude between two interacting objects, the effects can be dramatically different. When a gun is fired, the force exerted on the bullet by the gun is exactly equal in magnitude (but opposite in direction) to the force exerted on the gun by the bullet. Since the gun is much more massive than the bullet, the acceleration of the gun is much smaller than that of the bullet ($\mathbf{F}/m=\mathbf{a}$, thus a larger m produces a smaller $\mathbf{a}$ for a fixed $\mathbf{F}$). Thus we see why the change in the motion of the bullet is huge compared to the change in motion of the gun.

Momentum

The effects of objects on one another due to Newton's Third Law can be better understood in terms of another concept, **momentum**, which is defined as the product of the mass, m, of an object times its velocity, $\mathbf{v}$, therefore momentum$=m\mathbf{v}$. Momentum is a vector quantity, having a magnitude and a direction.

When the resultant external force on an object or group of objects is zero, the net acceleration of the objects is also zero, since from Newton's Second Law it equals the resultant force divided by the mass. But a zero acceleration means that the net velocity remains constant in time. Therefore, the momentum remains constant in time. This can be summarized, as follows:

Momentum is constant in time in a system on which there are no *net* external forces.

This concept is particularly important in understanding what happens during the firing of a rocket or gun. Let M be the mass of a gun and m be the mass of a bullet that it fires (see Fig. 2-6). Before the bullet is fired, the momentum of the gun plus bullet is zero since both have zero velocity. After the bullet is fired, the momentum of the gun and the bullet remains zero. Thus $0=mV-Mv$.

Exactly the same principle is involved in all rockets. Hot gases of mass, m, are exhausted at high speed V out of the rear of the rocket. The momentum mV of these hot gases being emitted backwards causes the main body, M, of the rocket to move forward with the velocity, v, such that $0=Mv-mV$ once again.

Let's try applying these principles to science fiction films by considering the film *The Day the Earth Caught Fire*. In that film, the simultaneous detonation of two hydrogen bombs (by the US and the USSR) supposedly pushes the Earth toward the Sun (74 min). The mass of the Earth is about 6×10^{24} kg, which is equivalent to the mass of 6.6×10^{21} tons of matter. The largest hydrogen bomb which has been exploded (thus far) released energy equivalent to an explosion of 68 million tons of TNT. We are not given the magnitude of the explosions generated by the two bombs in the film. Thus, we shall have to estimate the effect of an explosion much greater than that which has occurred to date.

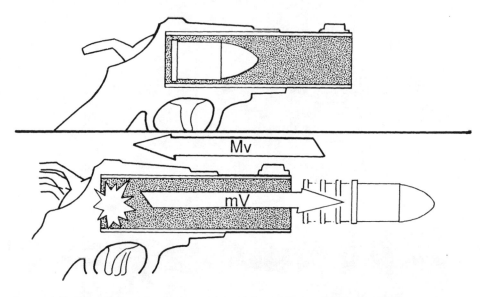

FIGURE 2-6. The backward momentum Mv of the gun is equal in magnitude to the forward momentum mV of the bullet.

For the Earth to be moved toward the Sun, some of its constituent molecules would have to be propelled in a direction away from the Sun at speeds greater than 25,000 mph, so as to escape the gravitational attraction of the Earth. At lesser speeds the molecules would eventually return to the Earth, which would not then experience a net velocity toward the Sun.

Let us assume that the explosions were sufficient to hurl 10^8 tons of debris at 25,000 mph. This would take an explosion immensely greater than any hydrogen bomb exploded to date since, as we know, it took huge multistage rockets to accelerate the small spaceships carrying our Apollo astronauts to comparable speeds. From the conservation of linear momentum: $0 = (10^8 \text{ tons} \times 25,000 \text{ mph}) - (6 \times 10^{21} \text{ tons} \times V_{mph})$; solving for V, we get $V = 4.2 \times 10^{-10}$ mph. Converting this into the distance traveled per year shows that the Earth would have moved only about 3.7×10^{-6} miles (or about one-quarter in.) closer to the Sun in 1 year. Since the average distance of the Earth from the Sun is 93×10^6 miles, this motion would not have been noticeable. It would not have dramatically raised the temperature of the Earth's surface as depicted in the film.

You may think that we have erred in not making the explosions even larger. However, as the explosive power of the hydrogen bombs is increased in the calculation, there is the likelihood that the detonation itself would cause world-wide destruction. In fact, any explosion too small to destroy all life on Earth either immediately or as a consequence of throwing tremendous dust clouds around the planet, would be too small to move the Earth appreciably toward the Sun. Our

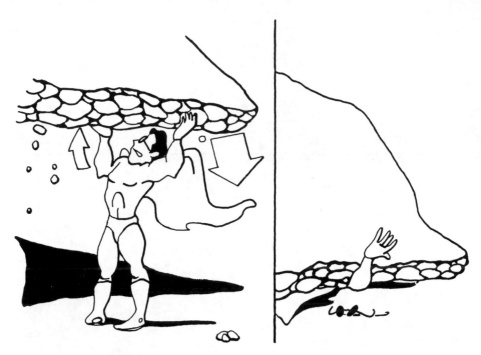

FIGURE 2-7. A very strong physics professor tries unsuccessfully to lift a segment of rock along the San Andreas fault.

calculation also neglected the fact that the Earth turns on its axis once every 24 h and is moving in a nearly circular orbit around the Sun. Including these rotational effects would make the calculation more complicated but not change the conclusion.

Another film which violates Newton's Third Law and the conservation of momentum is *Superman*. Near the end of the film Superman struggles to overcome the effects of an atomic explosion which has triggered violent motions along the San Andreas fault in California (111 min). In order to reduce the effects of this movement of the fault, Superman lifts a long segment of rock back into place as depicted in Fig. 2-7. The rock would have weighed hundreds of thousands, if not millions, of tons. If Superman exerts a force upward on the rocks, then by Newton's Third Law the rock slab would exert a force downward on Superman and push him through the rocks on which he was standing. It is important to realize that it does not matter how strong Superman is: the rocks on which his feet are standing would be unable to support a weight of millions of tons of rocks pressing down on him. Superman would be pushed into the Earth. One may also consider this example from the point of view of the conservation of momentum. If Superman moved a huge mass M up at a speed v, he would be pushed downward into the Earth at a much greater speed V.

A film which accurately portrays conservation of momentum is *2010*, in which two astronauts are depicted transferring between a Russian spacecraft, the *Leonov*, and an American spacecraft, *Discovery* (43 min). The Russian crewman uses a cylinder of compressed air as the propellant system: he fires the air toward the Russian spacecraft and the two astronauts are propelled in the opposite direction toward the American spacecraft.

Newton's Second Law may be rewritten to provide greater insight into nature. $F = ma = \Delta mv / \Delta t$, thus multiplying both sides of the equation by Δt yields

$$\text{force} \times \text{time interval} = \text{change in (mass} \times \text{velocity)}.$$

Alternatively, using the term **impulse** $=$ force $\times$ time and momentum $=$ mass $\times$ velocity,

$$\text{impulse} = \text{change in momentum}.$$

In this form it is easier to use Newton's Second Law to understand a wide variety of phenomena. For example, how does a karate expert break a concrete block? The block is broken by the force applied to it. That force comes from the hand of the karate expert which hits the block, resulting in a change of momentum of the hand that is equal to the total momentum of the hand just before it strikes the block. Thus, the hand should be moving as fast as possible before striking the block. In addition, to maximize the force, the time interval over which the momentum changes should be minimized. This is why some karate experts have a hard callous along the edge of their hands. The hard callous does not "give" the way soft skin would when delivering the blow; the delivery time of the blow is shortened and the applied force maximized.

Another example is learning how to land when jumping from a considerable height, as in sky diving. Here you wish to minimize the force on your body for a given change in momentum. Hence you want to increase the time interval over which the momentum changes. This is achieved by bending the knees upon making contact with the ground, thereby extending by a factor of 20 or more the time during which the jumper's momentum is being reduced compared to the time interval if the jumper lands stiff-legged. That reduces the force received by the jumper by a factor of 20 or more and may be the difference between walking away from the jump or being carried away with broken bones. Similarly, a person is better off jumping onto a wooden floor rather than a concrete floor, since the wooden floor *gives* somewhat, thus extending the time of impact and reducing the force of impact.

In other situations, the force may be relatively constant and the goal may be to maximize the change in momentum by maximizing the length of time over which the force acts. This is the case when a baseball player hits a ball. The player "follows through" with the bat to keep it in contact with the ball for as long as possible, thereby maximizing the impulse given to the ball and thus the change in momentum of the ball. This, in turn, maximizes the distance the ball will travel before striking the ground.

Energy

Probably the most fundamental concept in all of science is that of energy. Energy can take many different forms. It can appear as the motion of matter, as heat, as light, as an electrical current, or it can be released by a chemical reaction or a nuclear or thermonuclear reaction.

Let us start by considering one form of energy, **work**. When we lift something against Earth's gravity, work is done. The heavier the object or the higher we lift it, the more work is done. Work is defined to be force×distance the force moves. In symbols,

$$W = Fd,$$

where W is the work done by a force of magnitude F which moves something a distance d. The unit of work (or of energy generally) in the SI system is the joule (J), where 1J equals 1 N moving an object a distance of 1 m. In the British system the unit of work is the foot-pound since force is measured in pounds and distances in feet.

Note that by this definition no work is done if you stand motionless holding 200 lb. Your muscles will get tired, but no work has been done since the weight has not moved. If you lift a given weight twice the distance, you do twice as much work. Similarly, if you lift twice the weight a given distance you do twice the work.

The definition of work says nothing about how long it takes to do the work. We need to define another concept, power, to determine this. **Power** is defined as work divided by the time it takes to do the work.

$$P = W/t,$$

where P is the power in watts, if W is the work in joules and t is the elapsed time in seconds.

One often deals with multiples of the basic power unit: 1 kW is 10^3 W and 1 MW is 10^6 W. In the British system the basic unit of power is the horsepower, which is equal to 550 ft-lb/s or 746 W.

Suppose you wish to calculate how long it will take you to lift 1000 kg of water 5 m using a 1 kW motor. As we shall see in the section on gravity, one 1 kg weighs about 10 N on the surface of the earth. Hence, to lift 1,000 kg of water would require a force of 10,000 N (equal to the weight of 1,000 kg) and work of 10,000 N×5 m=50,000 J. The motor can provide 1,000 J/s, and thus the time required to lift the water would be at least $(50,000\ J)/(1,000\ J/s)=50$ s. The actual time might be longer since some of the motor's energy might be used up in overcoming internal friction.

When work is applied to an object it is converted into kinetic energy KE, potential energy, or heat (due to friction). **Kinetic energy** is the energy an object has due to its motion. If m is the mass of an object moving at speed v, the kinetic energy is given by:

$$KE = 1/2\ mv^2.$$

The kinetic energy of a moving body is equal to the work it can do while coming to rest. Let F be the magnitude of the force the body can exert in coming to rest and d the distance it travels,

$$Fd = 1/2\ mv^2.$$

Suppose we apply this equation to the problem of stopping a car. We want to determine how far a 1,000 kg car will travel after its brakes are applied. Assume that the brakes can apply a force of 5,000 N between the tires and the road: that is equal to about half the weight of the car and is a realistic assumption for a dry road and good tires. Suppose that the car is going 60 mph (or 26.8 m/s). Then d, the stopping distance, is equal to $mv^2/2F = (1000\ kg)\ (26.8\ m/s)^2/(2\times5,000\ N) = 71.8$ m or about 235 ft. Note that the stopping distance increases as the square of the speed of the car. If you double the speed of your car, it will travel four times as far after the brakes are applied. Thus, a car going at 30 mph on the same road would require a stopping distance of only $71.8/4 = 18$ m.

Potential energy is energy stored in an object in a variety of ways. The increase in energy caused by elevating an object is said to be due to an increase in its gravitational potential energy. In that case, the increase in potential energy is equal to the work required to lift the object to the elevated position.

Gravitational potential energy = weight × height moved.

Similarly, a compressed spring has potential energy equal to the work required to compress the spring. Gasoline has chemical potential energy stored in it. Potential energy is so named because it has the potential for doing work.

The most general law of physics refers to energy. It is called the **Law of Conservation of Energy**:

Energy cannot be created or destroyed but it may be transformed from one form to another. However, the total amount of energy never changes.

As an example, consider the experience of diving into a swimming pool, as pictured in Fig. 2-8. Let W be the weight of the diver and h the vertical distance from the diving board to the top of the water. The diver has potential energy of Wh on the diving board, but zero kinetic energy at the instant at which he steps off the diving board. When he hits the water, his potential energy is zero, but his kinetic energy is now equal to Wh. In other words, the potential energy he had when stepping off the diving board has been converted into kinetic energy.

potential energy

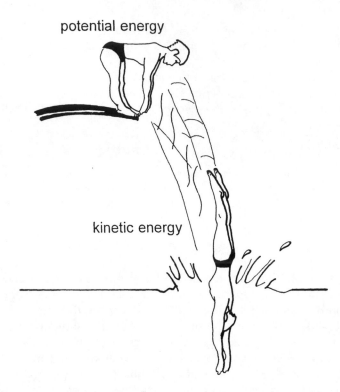

kinetic energy

FIGURE 2-8. A diver converts potential energy on the diving board into kinetic energy when striking the water.

As we shall see in the sections on nuclear physics and relativity, even matter is a form of energy and can be transformed into other forms of energy, as in a nuclear reactor.

We next apply the Conservation of Energy Principle to several science fiction films. We should note that perhaps the most common scientific mistake made in science fiction films is the violation of the Conservation of Energy Principle.

First let us consider the film, *Forbidden Planet*, in which a gigantic, 8,000-mile[3] machine built by an alien race, the Krell, on the planet Altair 4, has the ability to transform energy into matter—in any form. It is activated by the subconscious mind of Dr. Morbius, the sole survivor of a scientific expedition to the planet. Using the machine, Dr. Morbius creates a "monster from the id" which attacks members of a rescue expedition. The monster cannot be destroyed because the machine is recreating it every microsecond. While we do not presently have the ability to create matter from energy on any point on our planet, there is no violation of fundamental scientific principles in assuming that a more technologically advanced race might be able to do so.

The film correctly notes that it takes energy to create the monster. In fact, the dials on the wall of a Krell laboratory graphically depict the amount of energy being consumed in creating the monster. Since each dial on the wall of the lab reads 10 times the energy consumption of the preceding dial, one can estimate the relative power being drawn during the activities in which the film shows us the lab's dials. It should be noted that there are a total of perhaps 30 or 40 such dials. This represents a large power source, but it is not, as Dr. Morbius claims, "ten raised almost literally to the power infinity" (58 min). Ten to the 40th power is just as far from infinity as one!

Count how many dials are lit up when the monster launches an all-out attack on the rescue expedition (76 min). If you observe the dials carefully, you will see at least 16 of them are lit at the height of the attack (perhaps even more dials are lit to the right of the display seen in the film, but you can see 16 of them lit). We do not know the energy units used by this alien device, but we do see earlier in the film that only part of the first dial is lit when the Krell learning device is used. Thus the energy being used to create the monster from the id is at least 10^{15} times greater than the energy used to operate the Krell learning machine. If we were to assume that it took 100 W of power to operate the learning machine (the amount of power consumed in a 100 W bulb), then the monster was consuming $100 \times 10^{15} = 10^{17}$ W of power. How much power is this? There are about 10^8 households in the US; this would provide each household with 10^9, or 1 billion W, of power. That would light 10 million 100 W light bulbs in every household in the US!

The sources of all of this energy are 9,200 thermonuclear reactors. We do not know the power output of even one of these reactors. However, in order to provide at least 10^{17} W would require each reactor to provide $10^{17}/9{,}200 = 10^{13}$ W, or about 10,000,000 MW, far more energy than any human-built reactor can possibly produce. Since even more dials light up in the final sequence of the film in which the monster from the id tries to break into the Krell laboratory, the power output from each reactor must be even greater than that derived from the above calculation.

Doctor Morbius' robot, Robby, creates a variety of objects, including 100 yards^2 of lead shielding, 2 in. thick, for use by the rescue expedition (30 min). How do you think the robot achieved these creations? Since the mass of the many slabs of lead shielding clearly exceeds the mass of the robot itself, Robby could not have created the shielding using an internal power source. This is because any internal process would have to take part of Robby's mass and change it into other matter, but by the conservation of energy theorem (and remember that mass is a form of energy) Robby could not create more matter than he started with. The only plausible explanation is that the robot used the great Krell machine to create the huge slabs of lead, using the 9,200 thermonuclear reactors to supply the energy which was then transformed into matter.

There are other examples from more recent films, including *Star Wars* and *The Empire Strikes Back*. Both of these films repeatedly depict the ability of the Jedi Knights to move things with their minds. In one scene from *The Empire Strikes*

Back, the diminutive Jedi master, Yoda, lifts an entire spaceship just by using his mind (71 min). Since the hero, Luke Skywalker, seems to move about normally on the planet, his weight and hence the planet's gravitational attraction for objects on its surface must be similar to that on Earth. Thus, it is reasonable to conclude that the spaceship would weigh about the same as on Earth, namely, several tons at least. In order to move the spaceship upward, Yoda must apply a force at least as large as·its weight downward. This force applied upward will do work in moving the rocketship. Let us assume that the spaceship weighed 30,000 N (that is the weight of 3,000 kg on Earth and is equivalent to 3.3 tons). Then in raising it 5 m, Yoda performed $5 \times 30,000 = 150,000$ J of work. This is equivalent to the work done by a 100 kg adult (who weighs 220 lb on Earth) climbing 150 m (or about 450 ft) vertically upward! How could the creature's mind provide the energy to perform this massive amount of work?

Throughout both films there are illustrations of the Jedi Knight's ability to move things by looking at them. In the beginning of *The Empire Strikes Back* (9 min), Skywalker retrieves his light saber that is embedded in the ice, and at the end of the film Darth Vader hurls heavy objects at Skywalker (105 min). Both accomplish these feats with their minds alone. Yet the stories are inconsistent on the use of this power. Why waste time and energy hurling objects at one's opponent, or engage in hand-to-hand duels using light sabers? Why not simply snap the opponent's neck? It is a more effective way of resolving a one-on-one combat situation, but obviously was not used in the films because the duels would have ended rather quickly. This illustrates that the plots of the films are less rational than dramatic.

For those readers who are still skeptical about the statement that objects cannot be moved by an individual's mind because the power output of the mind is too small to move everyday objects, consider the following question: If someone had such powers, how would that individual use those powers to make money? Perform as a magician? It would be far easier and more profitable to go to gambling casinos. If a gambler can make the roulette ball drop in a given pocket and wager on the number of that pocket, the gambler will win 35 times the amount wagered. One could easily win thousands or tens of thousands of dollars per hour. Of course the casino managers would soon grow suspicious of such an individual and might have the player barred from playing roulette at that casino. However, has anyone ever heard of a player being barred from playing roulette because they win too frequently, indicating that they may have such mental powers?

Let us now consider another violation of the law of conservation of energy, this time using the box office blockbuster *Terminator 2: Judgment Day*. In the film the "heavy" is a robot from the future that is composed of a liquid metal that can transform itself into anyone/anything that it touches. What is the energy source for this creature? At one point in the film the creature has been separated (by being frozen and then shattered) into many small bits of metal. As the pieces warm up they **move** across the floor of a factory to join up with one another (120 min). But it takes energy to move the pieces—what is the power source? The film is silent on this point—and no obvious source can be deduced. All of the pieces move at the same time, not just one piece that might have contained an energy source. Nor is it plausible to suggest an energy source distributed throughout each molecule of the creature—we know of no mechanism for doing that. By contrast, the robot played by Arnold Schwarzenegger in the film has an internal energy source that will provide energy for over 100 years of operation. This does not violate any fundamental physical law.

Rotational Motion

An object that moves in a circle at constant speed v is said to experience **uniform circular motion**. The magnitude of the object's velocity remains constant, but its direction changes continuously as it moves around the circle. Since acceleration is defined as change in velocity divided by the time elapsed, a change in direction of velocity represents an acceleration, which we refer to as **centripetal acceleration**. It can be shown from trigonometry that an object moving in a circle of radius r at constant speed v has an acceleration a_c whose direction is toward the center of the circle and whose magnitude is

$$a_c = v^2/r.$$

The acceleration points toward the center of the circle, but the instantaneous velocity always points in a direction tangential to the circle. The units used must be consistent: thus a_c is in meters/second^2 if v is in meters/second and r is in meters.

Thus, it will require a net force to make an object of mass m move in a circle at constant speed. The equation derived for the centripetal acceleration provides the expression for a_c, but it does not provide any information about the force producing the acceleration. This force pointing to the center of the circular orbit may be due to gravity or to a rope joining the object to the center of a circle, or anything else.

Figure 2-9 pictures a ball of mass m being whirled at constant speed v in a circle of radius r by a cord. Using Newton's Second Law **F**=m**a** and the expression for **a** given above yields for the magnitude of the force on the ball $F=mv^2/r$.

Another example of centripetal acceleration occurs when a fast-moving car rounds a curve. As a passenger in the car, you may feel that you are being thrust outward, but this is because your body tends to move in a straight line whereas the car is following a curved path. To make your body go in a circle something must exert a force on it: the back of your seat, or your seat belt, or the door of the car. The inward force on the car is supplied by friction between the tires of the car and the road.

Consider the following example. A 1,000 kg car rounds a curve on a flat road of radius 50 m at a speed of 14 m/s. Will the car make the turn if friction can provide a force equal to half the weight of the car?

$$F=mv^2/r=(1{,}000 \text{ kg}) (14 \text{ m/s})^2/50 \text{ m}=3{,}900 \text{ N}.$$

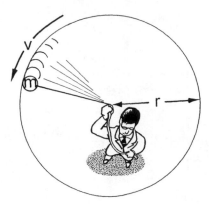

FIGURE 2-9. A ball of mass m, being whirled at constant speed v, in a circle of radius r, by the pull of a string. The force on the ball due to the string is directed towards the center of the circle.

The weight of the car is 10,000 N and half of that is 5,000 N, which is much larger than the 3,900 N needed to make the turn.

One may rewrite this equation in terms of the angular rotation of the object. The speed v of an object moving in a circle of radius r is given as $v = \omega r$, where ω is the angular velocity of the object with respect to the axis of rotation: ω has units of radians/second where 2π rad $= 360°$, or one complete revolution. Thus, one may rewrite the formula for the force on an object moving at speed v in a circle of radius r as

$$F = mv^2/r = m\omega^2 r.$$

An application of this centripetal force is to create artificial gravity on a space station by rotating it. This example also involves the rotational analogy to inertia: an object rotating about an axis will tend to remain in rotation about the same axis unless subjected to an external force. This property of an object is called **rotational inertia**. In order to quantify the effect of the external force, we introduce the term, **torque**, which is defined to be a force times the perpendicular distance between the line of action of the force and the axis of rotation.

An example will clarify the concept of a torque. Consider the motion of a door when you push on it. Figure 2-10 illustrates three places that you might apply a push to the door. The most effective place would be at the outer edge, labeled A. At point A the perpendicular distance (sometimes referred to as the lever arm of the force) between the line of action of the force (that is an imaginary line along which the force is applied) and the axis of rotation (the hinge) is largest and the push will be most effective. The same force applied midway along the door at B will have only half the lever arm as the force applied at A and will thus produce only half the torque: it will be only half as effective in rotating the door. Finally, a

FIGURE 2-10. A force applied at three points along a door produces three different torques about the hinges.

force applied at C, directly against the hinge, has a zero lever arm and will thus produce zero torque: the door will not rotate.

Torque explains why it is easier to turn a nut with a wrench that has a long handle than one with a shorter handle. The longer handle provides a longer lever arm for the force you are applying, and hence greater torque to turning the nut

Like inertia for linear motion, rotational inertia depends upon the mass of the rotating object. But it also depends upon the mass distribution of the object with respect to the axis of rotation. The greater the distance between the bulk of the object's mass and the axis of rotation, the greater the rotational inertia. This fact is used by tightrope walkers to help maintain their balance. These performers use long poles whose mass is mostly far from the axis of rotation, the tightrope. If the walker begins to lose his balance, his grip on the pole will start to rotate the pole. The rotational inertia of the pole resists, and the walker is able to regain his balance.

Let us return to a consideration of producing artificial gravity on a space station. Suppose that we start the station rotating. If it is shaped like a bicycle tire (see Fig. 2-11), anyone inside the "tire" would feel herself pressed against the outer wall of the structure, similar to the merry-go-round rider who feels pulled off of the platform. As stated earlier, the rider's body simply wants to move in a straight line whereas the space station forces the passenger to move in a circle by applying a force against the passenger's feet. Some amusement parks have cylindrical cham-

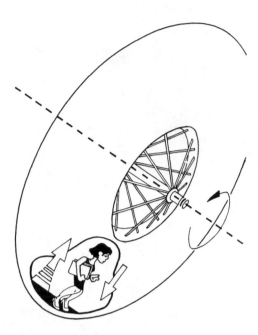

FIGURE 2-11. A passenger in a rotating space station would experience the sensation of being "pulled" against the outer wall of the station. This pull would result in an inward push by the wall against the feet of the passenger.

bers that rotate quickly, causing its occupants to be pressed against its walls. The floor of the chamber drops away, but the passengers do not fall down as long as the chamber continues to rotate sufficiently fast.

These examples are analogous to the effects of gravity on an individual. The Earth attracts the individual to the ground which pushes up against the person's feet to prevent them from sinking into the Earth. In the space station example, the walls of the station push against the feet of the individual. Since people find it difficult to adjust to rotational rates of greater than two or three rotations per minute, the space station would need to be huge to simulate an acceleration equal to that of gravity on the surface of the Earth—about 1 or 2 km in diameter!

The film *2001: A Space Odyssey* depicts such a giant rotating space station (20 min). The viewer sees both the outside and part of the inside of the station. The station is rotating at about two revolutions per minute: its size is huge compared to the space shuttle that lands on it. We do not know the exact size of the shuttle, although it must be at least 50 ft long, which would mean that the space station is perhaps 2,000 ft in diameter. If we use these figures, a station 2,000 ft (600 m) in diameter rotating at two revolutions per minute, then $r=300$ m and $\omega=0.2$ rad/s, the acceleration along the outer wall of the station will be about 12 m/s^2 or about 1.2 times the acceleration due to gravity at the surface of the Earth. Perhaps our estimation for the radius is incorrect: if the station is only 300 m in diameter, then the acceleration would be only 6 m/s ^{2}.

In the film *2010* we see another accurate portrayal of rotational motion. The two astronauts who transfer from the Soviet spacecraft, the *Leonov*, to the American spacecraft, *Discovery*, have to overcome the rapid rotation of the *Discovery* (46 min). It is rotating about a point at its center at the rate of about 3 revolutions/min. Given that it has an overall length of about 700 ft (or slightly more than 200 m), the distance from the center of the spacecraft to the bottom of the command module is about 100 m. The angular velocity is about 3 revolutions/min or 0.3 rad/s, and the maximum acceleration $[=(0.3)^2 100]$ experienced by the astronauts as they walk along the spine of the spacecraft to the command module occurs at the base of the command module. It is about 9 m/s^2 or about 0.9 that of gravity on the surface of the Earth. This is slightly less than that mentioned in the film which says the astronaut will "be in full gravity before he reaches the command module." Note that the American astronaut says "I'm getting heavy," but that is not what is actually happening. His weight is not increasing—it is rather an increase in the centripetal acceleration that he is experiencing as he walks away from the axis of rotation of *Discovery*.

Gravity

Sir Isaac Newton was the first to realize that the same force which caused objects to fall to the Earth was also responsible for the motion of the planets around the Sun and that of the Moon around the Earth. He stated that this force, gravity, attracted all matter in the universe to all other matter. The force of attraction lay along the straight line joining the masses. The magnitude of the force F of gravitational attraction between two mass points (depicted in Fig. 2-12) is given by

$$F = Gm_1m_2/r^2,$$

where m_1 and m_2 are mass points or the centers of spherically symmetric masses separated by a radial distance r. In the SI system of units, masses are in kilograms, r in meters, and G, the universal gravitational constant, is 6.67×10^{-11} N-m^2/kg^2.

Newton later invented calculus to prove that one could treat the mass of planets, which are extended bodies and not mass points, as though the mass of the planet was concentrated at its center, in order to calculate the effects of the planet's mass on objects located on or outside of the planet.

The force of gravitational attraction between two mass points decreases as one over the square of the distance $(1/d^2)$ between the points and is referred to as an "inverse square law" because of this spatial dependence. Thus, if you double the distance between two objects you decrease the gravitational force between them by a factor of $1/2^2$ or $1/4$.

It is the force of gravity that keeps the Moon and any artificial satellites in orbit about the Earth. We have already shown that it will take a force to put an object

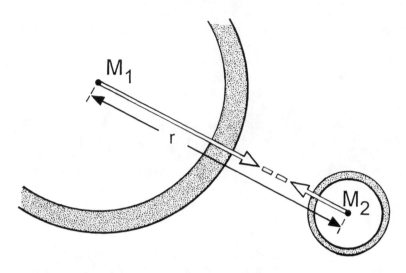

FIGURE 2-12. The force of gravity between two spherically symmetric masses, M_1 and M_2, acts along the straight line joining the centers of the masses.

into a circular orbit, because its velocity constantly changes and hence it has an acceleration. A force is needed to produce this acceleration: in the case of satellites moving in circular orbits about the Earth, that force is gravity.

Since the force of gravity on such a satellite points toward the center of the Earth, one may ask, why does the satellite not crash into the Earth? The answer is, why should it? Newton's Second Law states that the acceleration must be in the direction of the net force on an object, which is the case for an object in circular orbit about the Earth. No law of physics states that the velocity need be in the direction of the resultant force, only that the change in velocity (and hence the acceleration) must be in the direction of the resultant force.

For more complicated mass configurations than a sphere or a point mass, one would have to break up the masses into little spheres and then add all of the forces between these spheres in order to determine the resultant gravitational force between two irregularly shaped masses.

Next let us consider the gravitational force on a person of mass m standing on the surface of the Earth as shown in Fig. 2-13. From Newton's Law of Universal Gravitation, the force on the person's mass is GmM/r^2, where M is the mass of the Earth, which is about 6×10^{24} kg. The radius of the Earth, r, is 6.4×10^6 m. Substituting these numbers into the formula for the gravitational attraction experienced by the mass yields, $F = m \ (6.67 \times 10^{-11}) \ (6 \times 10^{24})/(6.4 \times 10^6)^2 = m(9.8 \ \text{m/s}^2)$. Thus the force of gravity due to the mass of the Earth produces an acceleration, g, on the person standing on the Earth's surface of $9.8 \ \text{m/s}^2 = 32 \ \text{ft/s}^2$.

88230

LIBRARY
DEXTER MUNICIPAL SCHOOLS
DEXTER, NEW MEXICO

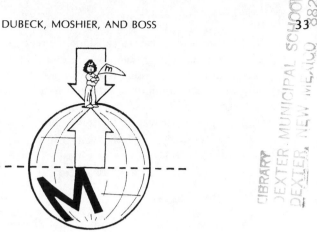

FIGURE 2-13. A person of mass m, standing on the surface of the Earth, is attracted to the center of the Earth by the mass of the entire planet M.

Historically, the preceding formula was used to determine the mass of the Earth from the experimentally determined value of the acceleration due to gravity on the surface of the Earth.

The acceleration due to gravity on the surface of any planet, $g', = GM'/(r')^2$, where M' is the mass of the planet and r' is its radius. Substituting the appropriate values for M' and r' yields Table 2.1 for the acceleration due to gravity for several members of our solar system.

TABLE 2.1. : Planetary Parameters

Body	Mass(kg)	Average radius (m)	$g'(m/s^2)$	g'/g
Moon	7.4×10^{22}	1.7×10^6	1.6	0.16
Venus	4.8×10^{24}	6.1×10^6	8.6	0.88
Earth	6.0×10^{24}	6.4×10^6	9.8	1.00
Mars	6.6×10^{23}	3.4×10^6	3.7	0.38

The last column of Table 2.1 gives the acceleration on the surface of the given body relative to that on Earth's surface.

On the Moon, which has a gravitational acceleration at its surface only 0.16 that on the Earth's surface, a man who weighed 200 lb on Earth would weigh $200 \times 0.16 = 32$ lb. This is because his weight is due to the gravitational attraction of the planet he is standing upon. On the Moon, the smaller gravitational attraction would result in the man weighing less. Note that the man's mass does not change in going from the Earth to the Moon, only his weight.

An object falling near the surface of the Earth has the same acceleration in the vertical direction irrespective of its mass. This acceleration is equal to $g = 9.8$ m/s². Thus an object dropped from rest has a speed at the end of a time t of gt. For example, at the end of 3 s the speed of the object is 9.8 m/s² $\times 3$ s $= 29.4$ m/s.

In the film, *The Black Hole*, the crew of a spacecraft approaching a giant black hole (which we will describe in the next chapter) finds that the gravity of the giant black hole **suddenly** increases substantially as the ship approaches it: this would not happen (6 min). The gravitational force of attraction between any two masses increases as one over the square of the distance between them and thus changes gradually, not in sudden jumps. Later the spacecraft enters a region near the black hole in which the gravitational forces disappear. Once again, this is science *fiction*: there is not any known way to neutralize gravity to produce a truly gravitational-free region of space. The attractive gravitational forces between matter extend throughout the universe.

Satellite Motion

The **escape velocity** (the minimum speed an object must have when leaving the surface of a planet so that its speed will never diminish to zero as it continues moving away from the planet) of the Moon is less than that of Earth because the gravitational attraction is lower. Thus it took less fuel for astronauts to take off from the Moon than from the Earth. The escape velocity of an object from the Moon is only 2.4×10^3 m/s as contrasted with the Earth's much larger escape velocity of 11.2×10^3 m/s.

Astronauts in orbit around the Earth experience the sensation of "free fall," i.e., the apparent absence of gravity. This does not mean that no gravity is being exerted upon the astronauts, but rather that the entire force of gravity is being used to pull the astronauts into a circular orbit about the Earth. There is no force of gravity left over to pull them into the Earth or to provide them with a sensation that gravity exists.

An easier example of this apparent absence of gravity is to imagine occupants in an elevator whose cable broke, and who are therefore falling to the bottom of the elevator shaft under the force of gravity. As Fig. 2-14 demonstrates, the passengers would not find that their feet were pressing down on the floor of the elevator since they and the elevator are both descending with the same acceleration—that of free fall at the surface of the Earth, 9.8 m/s^2. Thus if an elevator occupant had been standing on a scale, at the instant the elevator cable broke and it began accelerating downward, the scale reading of the passenger's weight would drop to zero. If the occupants were unable to see outside of the falling elevator they could not determine whether they were in free fall because the cable had broken, or whether gravity had magically "been turned off" (until, that is, the elevator hits the bottom of its shaft).

Returning to the concept of escape velocity, it should be noted that this applies to the molecules of the planet's atmosphere as well as to spacecraft. Thus, it is believed that the Moon lost its atmosphere shortly after its formation while Earth has continued to maintain its atmosphere over a period of at least four billion years.

Another way of looking at the escape velocity is to calculate the amount of work that must be done to lift a payload against the force of gravity to an infinite distance from the planet. The calculation is complicated by the fact that the force

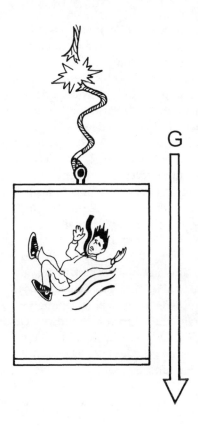

FIGURE 2-14. A person inside an elevator with snapped cables would be falling with an acceleration of 9.8 m/s².

of gravity on the object decreases as one over the square of the distance between the center of the planet and the escaping spacecraft. It turns out that 60×10^6 J of work are required per kilogram of a spacecraft to escape Earth's gravitational field at infinity. If one lets the initial kinetic energy equal this work, one calculates the initial velocity needed from KE $=(1/2)mv^2$, where KE $=60 \times 10^6$ J when $m=1$ kg, yielding $v=11.2 \times 10^3$ m/s.

If we provide any object with more than 60×10^6 J of kinetic energy per kilogram at the surface of the Earth, then, neglecting air resistance, the object will continue moving outward never to return to the Earth. Applying the same reasoning to the Sun, which has a mass 330,000 times larger than that of the Earth, one finds that even at a distance equalling Earth's orbit from the Sun, the escape velocity is 42.5×10^3 m/s. If we thus project a spacecraft at a speed greater than 11.2×10^3 m/s but less than 42.5×10^3 m/s it will escape the Earth but not the Sun. It will become another satellite of the Sun.

The first probe to leave the solar system, *Pioneer 10*, was launched from Earth in 1972 with a speed of only 15×10^3 m/s. It was directed towards Jupiter: its intentional near miss with that giant planet speeded it up sufficiently to exceed the escape velocity of the Sun at the distance of Jupiter. *Pioneer 10* contains information that will be of interest to extraterrestrials, if they should ever find it.

Artificial satellites are placed in orbit around the Earth for a variety of reasons, such as to relay television programs, provide weather information, or for military reconnaissance of other countries. For a satellite to go into orbit around the Earth it must be fired at a speed of between 8×10^3 and 11.2×10^3 m/s. At even greater speeds the satellite will recede indefinitely far from the Earth (if we neglect the attraction of the Sun). At speeds less than 8×10^3 m/s the satellite will fall back to the Earth because the Earth curves downward about 4.9 m for each 8×10^3 m of horizontal distance. Consider what happens to a satellite fired from a mountain at a speed of 8×10^3 m/s parallel to the surface of the Earth. If we neglect air resistance, the satellite encounters no force in the direction it is initially moving so that its horizontal velocity remains unchanged since no applied force means no acceleration. However in the vertical, or y, direction it experiences a net force, gravity. Gravity accelerates an object 9.8 m/s^2 near the surface of the Earth so that after 1 s its vertical speed is 9.8 m/s. Its average speed over the 1 s is the average of its initial and final speeds in the vertical direction which is $1/2(0+9.8)=4.9$ m/s. Hence in 1 s the distance it travels in the vertical direction=average vertical speed$\times$time=4.9 m/s$\times$1 s=4.9 m. But this is precisely how far the Earth has curved in the 8×10^3 m the missile has traveled in the horizontal direction. Thus, the satellite continues around the Earth in a circular orbit at a constant height above the Earth.

A satellite in circular orbit above the Earth always moves perpendicular to the Earth's gravitational pull. Thus, the force of gravity cannot speed up or slow down the satellite: no change in its speed occurs in the direction it is moving. A satellite in circular orbit thus continues to move at constant speed around the Earth. The crews of satellites in circular orbit feel weightless because all of the force of Earth's gravity is being used to pull them in a circular orbit: nothing is left to pull them towards the Earth.

In *2001: A Space Odyssey*, we see an accurate portrayal of the apparent weightlessness of a space shuttle approaching the rotating space station. A pen floats in the air (18 min) and a stewardess is seen seemingly walking "upside down" in the shuttle. But, of course, there is no "down" in the shuttle—down would be the direction of a gravitational force on the passengers, and they experience no such force. Later, we see that in the space station, as mentioned in the section on circular motion, the rotation of the station creates a centripetal force which feels like gravity to its occupants.

In *Aliens*, the starship carrying the troops to the planet (on which the alien creatures had been discovered in the film *Alien*) violates the law of gravity. We first see the starship moving in a straight line at constant speed (17 min), a motion consistent with Newton's First Law that an object on which there is no external force, including gravity, will move at a constant speed in a straight line. As the troops in the ship awaken from a state of hibernation, they move about as though they are subject to the normal gravitational field of the Earth. But there is no nearby planet to provide this gravity at this point in the picture. Nor is the ship rotating, which would produce an artificial gravity, as in *2010*. Furthermore, the soldiers are barefoot, instead of wearing special shoes which could adhere to the floor. They therefore should be floating about the starship.

When the troops later depart on a shuttle for the surface of the planet, the problem with the depiction of gravity is repeated. Presumably, the starship is in orbit around the planet, and in that case all of the gravity would have been used keeping the ship and its passengers in orbit and there would be nothing left over to provide the normal gravity the ship's passengers seem to experience.

Finally, the shuttle is depicted as "dropping" down from a bay hangar of the starship onto the planet (30 min). How is this possible? Any object that is part of the starship will have the same velocity as the starship relative to the planet. Thus an astronaut in orbit around the Earth does not plummet to Earth by stepping off of a space shuttle. We have seen astronauts take "space walks" around their spaceships. The astronaut and spaceship continue moving together in orbit around the Earth.

The shuttle in *Aliens* would have had to fire its rockets to move backward relative to the starship in order that its velocity relative to the planet was substantially reduced or made zero: only then would it fall towards the planet as depicted. But no such rockets are depicted as firing.

Exercises

1. *Motion*: Convert 30 m/s to mph.

2. Suppose a car, moving at 60 mph, comes to rest in 5 s. What was the average deceleration of the car?

3. *Newton's Laws of Motion*: Suppose that two forces, one of 10 N and the other 30 N, act in opposite directions on an object of mass 2 kg. What acceleration will the object experience?

4. A car and a fly have a head-on collision. The fly is splattered over the windshield. Is the force the fly exerts on the car smaller, the same, or larger than the force the car exerts on the fly? Is the acceleration of the car due to the impact smaller, the same, or larger than the acceleration suffered by the fly?

5. If you exert a force on an object and it does not move, what conclusion can you draw?

6. How much tension must a rope withstand if it is to accelerate a 1400 kg car at 0.5 m/s^2? Assume no friction with the road.

7. *Momentum*, Suppose that Superman stands between (and on) two railroad cars of identical mass. Is it possible for him to give either of the cars a greater push before he falls to the ground?

8. How does a "recoil-less rifle" (or bazooka) work? (It fires a shell without itself recoiling.)

9. Why does a roadway barrier made of barrels strapped together provide a better way of stopping a car that might strike it than does a concrete barrier?

10. Why are padded dashboards safer in automobiles?

11. *Energy*: How much work must be done in stopping a 3,000 kg truck moving 60 mph?

12. Suppose a car, previously at rest, rolls forward (with its engine off) from rest down a hill that is 100 ft above the floor of a valley. Neglecting friction with the road, what is the maximum height of the hill on the other side of the valley that the car could roll up without its engine being turned on?

13. If a car generates 20 hp when traveling at a steady 100 km/h, what must be the average force exerted on the car due to friction?

14. An inventor claims to have built a device that will give out a steady power of 100 W forever, without hooking it into an external power source. What do you think of the invention?

15. *Rotational Motion*: Do animals with shorter legs or longer legs have greater rotational inertia? Which animals run with a faster gait?

16. Suppose that the spacecraft *Discovery* in *2010* was rotating twice as fast. What would be the acceleration experienced by the astronauts when they reached the command module (assume the dimensions used in the textbook)?

17. *Gravity*: How much would you weigh on Mars?

18. Calculate the force of attraction between a 100 kg individual and a battleship of mass 50×10^6 kg, located at an average distance of 100 m from the individual. Would this force be noticeable?

19. *Satellite Motion*: Is the following explanation valid? Satellites remain in orbit instead of falling to Earth because they are beyond the pull of Earth's gravity.

20. Does the speed of a falling object depend on its mass? Does the speed of a satellite in orbit depend upon its mass?

Chapter Three
Astronomy

The Solar System: The Planets

Tycho Brahe (1546–1601) recorded the most accurate pretelescopic observations ever made of the positions of the planets as functions of time in a laboratory built for him by the King of Denmark in 1582. Brahe's assistant, Johannes Kepler (1571–1630), completed the analysis of these observations and consolidated them into three laws of planetary motion.

Kepler's First Law of planetary motion states that:
Each planet moves in an elliptical orbit around the Sun with the Sun occupying one of the two foci of the ellipse.

Kepler's Second Law of planetary motion states that:
The speed of each planet varies in such a way that an imaginary line joining the planet to the Sun sweeps out equal areas in equal times.

Kepler's Third Law of planetary motion states that:
The cubes of the semimajor axes of the elliptical orbits are proportional to the squares of the times it takes for the planets to make complete revolutions about the Sun.

Thus the orbits of the planets are ellipses rather than circles. As pictured in Fig. 3-1, when a planet is closer to the Sun, an imaginary line joining it to the Sun must still sweep out the same area per unit of time as when it is further away from the Sun. Thus, the planet must be moving faster when it is closer to the Sun than when it is further away from the Sun. Also, planets whose orbits are further away from the Sun take longer to revolve around it (this is called the period of their orbit).

Kepler's laws of motion describe how any two bodies which are bound together by their gravitational attraction move, whether they are in our solar system or not. It remained for Isaac Newton (1642–1727) to explain the basis for Kepler's laws in terms of his Laws of Mechanics and his Theory of Universal Gravitation. All three of Kepler's laws can be derived from Newton's laws using fairly simple mathematics.

Table 3.1 presents selected data on the Sun and the nine planets in our solar system. Astronomers have generally grouped the four inner planets–Mercury, Venus, Earth, and Mars–under the term "terrestrial planets." They are all relatively small and have a relatively high average density similar to that of the Earth (about 5.5 g/cm^3, which is 5.5 times the density of water).

Mars is probably the planet which has most fascinated its human observers. Many science fiction films and stories have attributed intelligent life to Mars. The American astronomer Percival Lovell (1855–1916) and the Italian astronomer Giovanni Schiaparelli (1835–1910) both reported seeing "canals" in their telescopic studies of Mars. The existence of these straight-line canals criss-crossing Mars

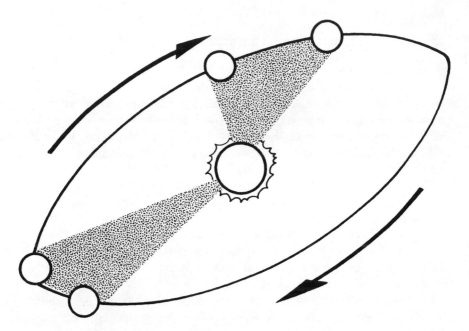

FIGURE 3-1. A planet moves in an ellipse about the Sun. The areas of the two shaded triangles are equal.

would have been strong evidence that Mars had at one time been inhabited by a technologically advanced civilization. But the close fly-bys of unmanned probes and the landing on the surface of Mars by the unmanned *Viking 1* and *Viking 2* found no evidence of such canals. Mars has a gravity at its surface about 38% that

TABLE 3.1. : Planetary Parameters

Planet	Average radius of orbit (km)	Mean diameter (km)	Mass/Mass Earth	Period
Sun		1,390,000	333,000	
Mercury	58,000,000	4,900	0.055	88 days
Venus	108,000,000	12,100	0.82	225 days
Earth	150,000,000	12,800	1.0	365 days
Mars	228,000,000	6,800	0.11	687 days
Jupiter	778,000,000	143,000	318	12 yr
Saturn	1,430,000,000	121,000	95	29 yr
Uranus	2,870,000,000	53,000	15	84 yr
Neptune	4,500,000,000	50,000	17	165 yr
Pluto	5,900,000,000	3,000	0.02	248 yr

at Earth's surface. It has lost most of its atmosphere, which has a pressure of only about 0.7% of Earth's sea-level atmosphere. Mars' atmosphere is 95% carbon dioxide with about 3% nitrogen, 2% argon, and trace amounts of oxygen and other elements. It is believed that water played an important role in Mars' history, carving out large canyons that are up to 2,500 miles in length. The Martian terrain contains craters from impacting meteors and also the remnants of volcanoes. The northern hemisphere is generally depressed with few craters. Its crust has been altered by volcanic activity and the associated lava flooding. The southern hemisphere has a densely cratered and elevated crust that has not changed appreciably throughout its history.

There is believed to be extensive water at both of the poles of Mars as well as in the surface of Mars down to a depth of several meters. Temperatures vary at the equator from 30 °C (about 86 °F) down to -130 °C. It is even colder over the polar regions. Wind speeds as high as 250 mph have been observed at heights of several miles above the surface. These winds are strong enough to create major dust storms.

Jupiter, Saturn, Uranus, and Neptune have been referred to as gaseous giants. They are all much larger than the inner four planets and have a much lower average density, on the order of 1 g/cm^3, the same density as water. The outer planets also have many more satellites than the inner planets (Earth has one, Mars two, and Mercury and Venus none). The gaseous giants also all have rings about them.

Jupiter has an unusual feature: the Great Red Spot, first observed three centuries ago. It is a giant cyclonic region of whirling gas that measures about 8,000 by 24,000 miles.

The satellites of the gas giants come in a variety of sizes and surface features. Io, one of the moons of Jupiter, has a mass 1.21 times that of Earth's Moon and is the most geologically active satellite in the solar system; it has active volcanoes and its orbit around Jupiter is inside of a torus (a doughnut-shaped cloud) of sulphur and oxygen atoms, which may have been emitted from Io's volcanoes in the past.

Much less is known about distant Pluto. Its orbit is the most eccentric, which means that it deviates most from a circle of any of the planets. Its orbit is so eccentric, in fact, that part of it lies inside the orbit of Neptune. Pluto is even smaller than our Moon and has at least one known satellite, Charon.

The Sun

The Sun is the source of light and most of the energy which we utilize on the Earth. The Sun is powered by a thermonuclear reaction (see the chapter on nuclear physics) which converts hydrogen into helium with the release of great amounts of energy. Each second about 4.4 million tons of matter are being converted into energy in the Sun. Some of this energy strikes the Earth and provides an essential ingredient for the growth of plants. The Sun has been providing this energy for at least 5 billion years, since the time when the gravitational contraction of its mass raised its temperature sufficiently to ignite the thermonuclear process.

The surface of the Sun is at a temperature of about 10,000 °F. Both temperature and pressure increase below the surface of this massive body. At its center the temperature is estimated to be about 20,000,000 °F, sufficiently high to permit the thermonuclear process to proceed.

The Sun is 330,000 times more massive than the Earth: it is about 860,000 miles in diameter. Thus, its average density is about 1.41 g/cm^3, about the same as the planetary gaseous giants. Because of its huge mass, which is mainly hydrogen and helium, it is believed that the Sun will be able to continue delivering energy at its present level for approximately another five billion years.

About 5 billion years from now, the Sun's energy output will decrease due to exhaustion of the hydrogen in its core, and it will collapse inward. This collapse will increase its internal temperature until unprocessed hydrogen outside its core is fused into helium. Over the next billion or so years the Sun will expand to a giant red star about the size of the orbit of Mercury; at this time its greatly increased energy output would raise the temperature on the Earth to the point where all life would be destroyed unless we have progressed technologically to the point that we can overcome even such a cosmic calamity. The Sun will pass through further changes over the next billion years or so until all of its fuel is exhausted; at that time it will become a small, dense dying star called a white dwarf.

The Moon

The Moon is our nearest planetary neighbor. It is 240,000 miles distant and is 2,160 miles in diameter. The acceleration of gravity on the surface of the Moon is only 16% of that on Earth's surface. This caused the Moon to lose its atmosphere soon after it formed because the escape velocity of molecules from the Moon was too low relative to the average speed of the molecules in its atmosphere. Their speed was determined by the temperature of the Moon's surface which would have received about the same energy per unit surface area from the Sun that the Earth received.

Without an atmosphere to burn up incoming meteorites, the surface of the Moon has been bombarded for four billion or so years by a variety of meteorites, from giant ones that left large craters to microscopic ones that caused the Moon's surface to be pulverized into a powdery substance called the regolith. All of the Moon's rocks are igneous, meaning they were formed by the cooling of lava. Analysis of the rocks collected from the Moon by our six manned Apollo missions (1969–1972) as well as by three unmanned Soviet Luna missions indicates that the surface of the Moon last cooled between 4.4 and 3.1 billion years ago. Thus the Moon is at least 4.4 billion years old. Present theories suggest that the Moon formed about 4.6 billion years ago and that the outer layer of the Moon remained molten for perhaps 200 million years before it cooled. Then the Moon was bombarded by large meteorites for a period starting at 4.2 billion years ago and ending

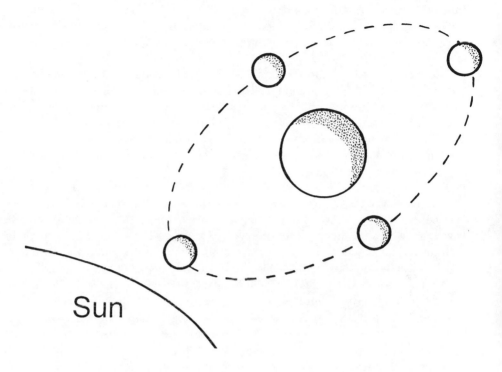

FIGURE 3-2. The side of the Moon facing the Sun is illuminated as the Moon moves in a nearly circular orbit about the Earth.

about 3.9 billion years ago. At that time the interior of the Moon was sufficiently heated (from radioactive elements inside it) for volcanoes to erupt. This vulcanism stopped about 3.1 billion years ago. Until then, the histories of the Earth and the Moon were very similar. However, Earth continues to be geologically active, and has also retained its atmosphere. These two facts have led to life on Earth compared to a lifeless Moon. It is not certain whether the Moon has a molten core as does the Earth.

The Moon moves in a circular orbit about the Earth. Since its period of rotation is identical to its period of revolution about the Earth, the same side of the Moon always faces the Earth. The Moon is illuminated by reflected light from the Sun, as indicated in Fig. 3-2. When the Moon is on the other side of the Sun from the Earth we see the illuminated side: this is called the full moon. When the Moon is between us and the Sun–the time of a new moon–we do not see the Moon at all.

A special case of the Moon being between the Earth and the Sun occurs when the Moon blocks the Sun's rays from reaching some area of the Earth; this is called an eclipse of the Sun. Since one can accurately predict the motion of the Sun,

Earth, and Moon using Newton's Laws of Motion and his theory of gravitation, one can accurately predict solar eclipses far into the future. One eclipse in which "totality" was achieved (i.e., a complete blocking of the Sun's rays by the Moon) over a substantial portion of the Earth occurred in July, 1991. Scientists recorded the eclipse to measure properties of the Sun such as the light coming from its outer layers.

The tides on the Earth are caused by the gravitational pull of the Moon and the Sun on water on the Earth's surface. The pull of the Moon on the water is about twice that of the Sun, which while having a much greater mass than the Moon, is also much further away. These planetary bodies cause two tidal bulges: one closest to the Moon and one on the opposite side of the Earth from the Moon. When the Sun and Moon are in line, as they are when there is a new or full Moon (see Fig. 3-2) the tides are the greatest.

The solar system has been the focus of many science fiction films. In *2010*, we are shown two of the Moons of Jupiter, Io and Europa. The spacecraft *Discovery* and *Leonov* are in orbit near Io and we see the effect this orbit has had on *Discovery* which has been there for 9 years (41 min). The spacecraft is covered with an orange powder, presumably a sulphur compound emitted by the many active volcanoes on Io, which itself is also accurately depicted as orange-colored.

Europa is portrayed in the film as the location on which a new life form has begun to develop. The *Leonov* detects the presence of chlorophyll, indicating that primitive life forms may already exist there (28 min). The surface of Europa is believed by scientists to be a crust of ice many miles thick; it has a complex array of cracks and streaks on it. No volcanic activity has been observed on Europa, however, and few impact craters are visible on its smooth surface. The picture of Europa presented in *2010* is consistent with these facts, although the ice is pictured as being only a few feet thick with the source of the chlorophyll located in a crater.

Jupiter and the Great Red Spot are seen in the background throughout the latter part of *2010*. At the conclusion of the film Jupiter is "ignited" and becomes another sun whose radiant energy will accelerate the development of intelligent life on Europa (105 min). Here the film conflicts with presently accepted theory that a mass larger than that of Jupiter is needed to provide sufficient internal gravitational pressure to raise the temperature high enough to ignite the thermonuclear process at the center of the mass. It is currently believed that the mass required to form a star is equivalent to at least 3,000 Earth masses, whereas Jupiter's mass is equal to only 318 Earth masses. In addition, if the thermonuclear process were initiated, the outward force of the reaction would have blown Jupiter apart and would have killed the occupants of the departing *Leonov* by either direct blast and heat or by lethal levels of radiation.

A minor error in the film is the glaringly bright light that is seen when the door of the *Leonov* is opened so that the astronauts may travel to *Discovery* (43 min). At Jupiter's distance from the Sun, the intensity of the Sun's rays would only be about 4% of those at the upper atmosphere of the Earth.

The film *Total Recall* has a novel ending, in which a power plant on Mars, built by the long-dead Martians, is activated (103 min). The power plant heats the planet's surface, releasing the trapped gases, and thus creates a breathable atmosphere for the entire planet. While it is believed that water may lie locked under the surface of Mars, heating water will provide water vapor, not breathable air. There might be some free oxygen locked in the ice with the water, but not nearly enough to form a breathable atmosphere.

Each water molecule consists of two hydrogen atoms and one oxygen atom; these molecules would have to be separated (this can be done by a process called electrolysis involving an electric current passing through the water) in order to get appreciable amounts of oxygen into the air. Even then, where is the nitrogen that comprises 80% of the air on Earth to be found? An atmosphere of only oxygen and hydrogen would be explosive.

The film also overlooks the vastness of Mars; it depicts the one power station as providing enough air to cover the planet. But, as Table 3.1 indicates, Mars is about half the diameter of Earth; its surface area is about 50 million miles^2! Its surface would need an atmosphere of density similar to that of Earth's: that much gas would not be available from the surface of Mars directly under a plant that was at most a fraction of a square mile in size. It would probably take many such power plants many years (or even centuries) to achieve an Earthlike atmosphere on Mars.

Finally, there is a logical inconsistency in the film: why would the Martians have gone to the trouble to construct such a plant (or series of plants) and then not have used it to preserve their atmosphere? The "heavy" in the film suggests that the Martians were afraid that activating the plant would cause a planet-wide melt-down. But would they then have constructed the plant in the first place and risk an accidental activation of the plant? Furthermore, faced with the depletion of their atmosphere would they have had much to lose by trying the process?

By contrast, the creation of a breathable atmosphere by a human-built plant in *Aliens* takes about 20 years (11 min).

Meteors and Comets

Meteors are small fragments of matter that approach the Earth as it moves in space. Usually we see meteors as bright streaks in the sky, sometimes called "shooting stars," which occur when they burn up due to friction with the atmosphere. Occasionally, large meteors are not completely burned up in the atmosphere and survive to strike the Earth. These are called meteorites.

Meteorites usually fall into one of three categories: stony meteorites with composition similar to ordinary rock; stony iron meteorites which are a matrix of stone and iron; and iron meteorites, consisting largely of iron with some nickel alloyed with it. The stony meteorites often contain organic compounds such as hydrocarbons and amino acids. These biologically important compounds evidently formed in the hot gaseous cloud (called a gaseous nebula) from which the solar system is believed to have formed. It is believed that meteorites (and their larger "cousins," asteroids) are leftover "building materials" from which the inner solar system was fabricated five billion years ago.

A large number of asteroids exist in our solar system. Some of them pass relatively close to Earth. However, the orbits of most of them lie between those of Mars and Jupiter. There are at least 100,000 asteroids that are a half mile or more in diameter. The largest, Ceres, is over 600 miles in diameter.

Comets also orbit our Sun. They are believed to be frozen icebergs consisting of a combination of frozen gases, such as ammonia and methane, with ice and meteor material trapped within them. Near the Sun, the solid gases evaporate forming the tail of the comet which is visible millions of miles away. The Sun's radiation pressure pushes the tail of the comet away from the Sun. When the comet passes the Sun, the comet freezes again into a small body only a few miles across. The most famous of the comets, Halley's Comet, is one of a class of long-period comets. It moves.in an elliptical orbit around the Sun once every 76 years.

We have crater evidence that at least 200 large meteors have struck Earth. It is estimated that at least one large collision occurs each 10,000 years. For example, near Winslow, Arizona, the Baringer meteorite crater is located. It is nearly 500 ft deep and 4,000 ft wide. It was created by a meteorite weighing at least 30,000 tons which struck Earth about 24,000 years ago. The collision must have devastated all plant and animal life in a large region around the crater.

Earth has had more recent collisions with large meteorites and/or comets. On June 30, 1908 a tremendous explosion occurred in a forested region near the Tunguska River in Siberia. The sight of a great ball of flame leaping from the forest was followed by an explosion powerful enough to level trees in an area of about 800 square miles. One estimate of the force of the explosion was that it was equivalent to 10–20 megatons of TNT exploding! No impact crater was located and thus the most probable cause of the explosion was a collision with a comet, which does not have a solid rock core.

Scientists have speculated that the dinosaurs may have become extinct because of a collision with an asteroid or comet. One theory suggests that every 26 million years or so a rain of comets that lasts for hundreds or thousands of centuries bombards the Earth. The impact of some of the larger asteroids or comets spews enough dust into the atmosphere so that it blocks the Sun's rays from reaching the Earth for months or years and most of the plant and animal species on Earth perish. According to this theory the last extinction occurred about 11 million years ago so that the next calamity should not occur for another 15 million years. Scientists are searching the skies for evidence of a possible "dark star," which they have named "Nemesis." They suggest that Nemesis may circle our solar system every 26 million years. When it approaches our solar system, Nemesis passes through the region of space beyond the orbit of Pluto which contains vast numbers of comets. Its passage would disrupt the comets' orbits and send a shower of them towards the Sun. Some of these comets would then collide with the Earth.

Most scientists are concerned that long before 15 million years have elapsed

Earth will be struck again by a single large asteroid or comet. The film *Meteor* vividly describes what could happen in this case. In the film, a comet entering the asteroid belt between Mars and Jupiter strikes one of the asteroids and sends a 5-mile-diameter object hurtling toward Earth at 30,000 mph (9 min). Scientists inform the authorities of the effects of such a collision. The picture shows the effects of Earth's collision with "splinters" of the struck asteroid. One splinter sends a 100 ft high tidal wave sweeping across Hong Kong and another has a direct hit on New York City; its impact is similar to a hydrogen bomb exploding over the city. The asteroid itself is prevented from striking the Earth by destroying it with dozens of missiles carrying hydrogen bombs.

Let us consider the energy that would be released in a collision with a 5-mile-wide asteroid moving 30,000 mph relative to the Earth. The kinetic energy of the asteroid is $mv^2/2$ where m is the mass of the asteroid. In order to estimate the mass of the asteroid, we will assume it is stony with the same density as the surface of the Earth, about 3.5 g/cm^3. This yields a mass of 9×10^{14} kg. Using the conversion factor 1 mph $= 0.447$ m/s, the speed of the asteroid becomes 1.3×10^4 m/s and it has a kinetic energy of 7.6×10^{22} J. This is enough energy to light a 100-W bulb for every man, woman, and child in the US for 100,000 years! The force of the impact would totally destroy a huge area around the point of impact and the dust thrown up by the impact could have global effects on the human race by significantly reducing the amount of sunlight that reached the surface of the Earth.

The Galaxy and the Universe

In order to describe the universe beyond our solar system we must introduce a new unit of distance, the **light-year**. This is the distance that light travels in one year. It is approximately equal to 5.88×10^{12} miles. In terms of the time for light to travel a given distance, the Moon is 1.3 light-seconds distant and the Sun is 8.3 light-minutes distant from Earth. By contrast the next closest star to Earth, Alpha Centauri, is 4.3 light-years distant. Thus, when you look at Alpha Centauri you are seeing it the way it was 4.3 years ago, when the light that is now reaching your eyes was emitted from the star. Our galaxy (or star island) is called the Milky Way and contains about 2×10^{11} stars. It is disk shaped, like a pancake, with a central bulge. The diameter of the disk is about 100,000 light-years and its thickness is about 3,000 light-years. The diameter of the central bulge is about 13,000 light-years and the thickness of the central bulge is about 10,000 light-years.

Our Sun is located about 30,000 light-years from the center of the Milky Way and is moving in an almost circular orbit through the galaxy with a period of revolution of about 225 million years. The Milky Way is itself rotating: the central region of our galaxy rotates like a solid body while the stars further out from the center, such as our Sun, follow nearly circular orbits. The mass of the galaxy is estimated to be on the order of 1.7×10^{11} times the mass of our Sun. This is consistent with the estimate that there are 2×10^{11} stars in our galaxy since about 95% of the mass of the galaxy may be contained in its stars. The remaining 5% of its mass is in gas and dust.

The Milky Way is a spiral galaxy, one of the most common types of galaxies. Galaxies often cluster in groups; there are over 20 galaxies in our particular cluster. The distances to other galaxies are enormous. For example, the distance to one of our nearest neighbors, the Andromeda Galaxy, is about 2×10^6 light-years. The furthest galaxies that we can see are on the order of 10^{10} light-years away.

Scientists believe that the universe is expanding and that the galaxies furthest from us are moving the fastest. This conclusion is based upon the analysis of the light reaching us from those galaxies. The light is shifted towards longer wavelengths (the so-called "redshift") as would be expected if the source of the light were moving away from us (see the Doppler Effect in Chap. 9).

Among the more interesting astronomical phenomena are the supernova, the pulsar, and the black hole. As a star of mass greater than 1.4 times that of our Sun ages and its hydrogen fuel nears exhaustion, it contracts until the density of its core becomes so great that it acts like a compressed spring and rebounds with a tremendous explosion. This explosion lasts but a few minutes and is called a supernova. In 1054, Chinese astronomers recorded a supernova explosion that was so bright that it could be seen by daylight. The glowing plasma remnants of that supernova now make up the spectacular Crab Nebula.

The evolution of a star after a supernova explosion will depend upon the mass of the star. Stars whose remnant masses are less than about three times the mass of the Sun will become either a white dwarf star or a neutron star. The radius of a neutron star will be on the order of a few miles. Its density will be that of the nucleus of an atom. These neutron stars may rotate rapidly; if they do so they are called pulsars since they emit pulses of low frequency radio waves with each rotation. These emissions can be detected with radio telescopes which detect radio waves in the same way that ordinary telescopes detect light.

When a star whose remnant mass is greater than three times the mass of the Sun collapses, its density and gravitational field are so strong that nothing can escape its attraction. It is called a black hole and is very small. We cannot see a black hole because even light is trapped by it. One way to understand black holes is to use the concept of escape velocity. For a black hole, the gravitational fields are so strong that the escape velocity exceeds the speed of light. Since according to Einstein's theory of relativity, nothing can travel faster than the speed of light, nothing escapes from the black hole. The distance from the black hole at which nothing can escape is called the event horizon. A star with a remnant mass of three solar masses would have an event horizon of about 9 km. This distance increases linearly with the mass of the collapsing star. Thus a star of six solar masses remaining in it would have an event horizon of 18 km.

Many strange things are expected to occur within the event horizon of a black hole, such as the slowing down of time. At the center of the black hole there exists a singularity, a point at which the mass of the collapsed star is concentrated into

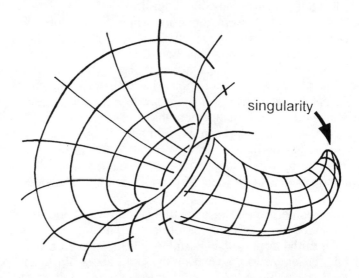

singularity

FIGURE 3-3. A graphical depiction of the distortion of space around a black hole. A singularity is believed to exist at the bottom of the cornucopia-shaped tube.

zero volume and infinite density. Refer to Fig. 3-3 for a graphical depiction of a black hole. No one knows if such a singularity actually exists, and many scientists are uncomfortable whenever any mathematical formulation of a physical phenomenon leads to a singularity.

Black holes may be responsible for another phenomena called **quasars**, which are galaxies emitting up to a hundred times more light than an ordinary galaxy such as the Milky Way. So far the speculation that best fits the observational data proposes that the energy source is a giant black hole with perhaps a mass of a billion solar masses; the source of the energy emitted is fueled by the black hole capturing gas and even stars from its nearby surroundings.

At the end of the film *The Black Hole*, a small spaceship enters a black hole (86 min) and the occupants experience some kind of time warp after which they come out of the other end of the black hole unharmed. This could never happen as anything entering the black hole would be crushed, killing all of the occupants. It is not known where a black hole may lead, but only subatomic particles would survive a journey through it, not people and machines.

The film does portray the event horizon and the increasing gravitational fields using graphics, similar to Fig. 3-3.

The Evolution of the Universe

At present, the most widely accepted theory supposes that the universe has been expanding steadily since some initial explosion, 15 or 20 billion years ago.

This is called the "big bang" theory. In this theory the universe was concentrated in one large mass which erupted in a tremendous explosion, hurling matter outward. As this matter cooled, the gravitational attraction between the gas and dust molecules caused them to come together in clouds, called protosuns. As the gas in what is now our solar system contracted, about 10% of the matter remained outside of its protosun to form protoplanets. It is probable that these protoplanets were huge compared to their present sizes, while our Sun was still dark. When the Sun's contraction raised its temperature high enough to initiate the thermonuclear process, the cold protoplanets grew warm and the envelopes of gas about them boiled away. After perhaps a billion years or so the shrunken planets we know today emerged from the clouds that had enveloped them. Simultaneously, the cores of these planets underwent pronounced heating, partly due to radioactivity inside them.

Ultimately entire planets, including the Earth, melted and the iron and silicate components separated. This process formed Earth's present molten core and mantle. Then the Earth began to solidify and to develop its present crust. This occurred perhaps 4.5 billion years ago, making the Earth about 5.5 billion years old as a planet. Presumably this process was repeated in other star systems, as the stars in our galaxy and countless others were formed.

The **cosmological principle** is central to any consideration of the future of the universe. It says:

All observers see the universe in the same way, i.e., the universe is homogeneous (uniform) and isotropic (the same in all directions).

In considering the future of the universe, a key question is whether there is enough matter in the universe so that gravitational attraction will stop the present expansion of the galaxies and then cause them to contract into one huge mass (the "big crunch"), which will undergo another big bang, starting the formation of stars and planets all over again. The total time of each such cycle of the universe has been estimated to be between 40 and 100 billion years, if this cyclic theory is correct.

Alternatively, other scientists have suggested that the universe may go on expanding indefinitely, since there is not enough matter to stop the galaxies from moving outward. Other theories include the continuous creation of matter to feed a steady-state universe: the new matter created compensates for the thinning of matter due to the expansion of the universe. In this model, space expands exponentially with time toward infinity. It is clear that as we deal with the origin and the future of the universe, we are getting more remote from the experimental observations that form the basis of modern scientific theories. We should therefore be cautious in our assertions of what may happen billions of years in the future.

Intelligent Life Elsewhere

No treatment of the solar system and beyond would be complete without addressing the issue of the possible existence of other intelligent life forms in the universe. As described in the previous section, scientists believe that the formation of our solar system from a dust cloud was repeated throughout our galaxy and others. They believe that billions of other stars have solar systems revolving about them and that some of these planets are likely to be inhabited by intelligent beings. The problem is to estimate the number of intelligent civilizations coexisting with ours and then to estimate the possibility of our making contact with them.

We can, with reasonable certainty, state that there are no such alien civilizations in our solar system. We have walked on the Moon, our unmanned probes have landed on Mars and Venus, and other probes have flown by most of the other planets. No trace of any life, intelligent or primitive, has been found. The conditions are so severe on the surface of most of the planets that it is difficult to imagine that any life would arise on them. For example, the very dense atmosphere of Venus contains acid and its surface temperature exceeds 800 °F. The gas giants are so far from the Sun that their temperatures are far below the freezing temperature of water.

We must, therefore, look beyond our own solar system for other civilizations. We will confine our analysis to our galaxy. It contains about 200 billion stars and the distances to the next nearest galaxy are so much greater than the dimensions of our galaxy that it seems reasonable to assume that any civilization we might contact would originate from within our galaxy.

In determining the number of such intelligent civilizations coexisting with ours a number of factors must be taken into account. Scientists have suggested various formulas to estimate the values of each such factor. We next consider one such formula. $N=(A \times B \times C) \times (D \times E) \times (F \times L)$. $N=$ the number of intelligent civilizations in our galaxy that can communicate with us. $A=$ the average annual rate of star formation during the Milky Way's existence. This is estimated by dividing the 200 billion stars by the 10 billion years the galaxy has existed, yielding $A=20$ stars/yr. $B=$ the fraction of stars that have planetary systems and is estimated from nearby stars to be $B=0.5$. $C=$ the number of planets in each planetary system suitable for life. Such planets must lie within an ecoshell about the star: they must not be too close to the star but also not too far away. It is important that the temperature on the surface of the planet be moderate, preferably between the freezing and boiling points of water. Based on our solar system this equals one planet per solar system: $C=1$. Note that A, B, and C are estimations of astronomical factors dealing with the formation of solar systems. $D=$ the fraction of planets on which life actually does appear. Life is probable only on planets whose parent star is able to provide a constant energy supply for a long period of time—about 20% of the stars. Thus $D=0.2$. $E=$ the fraction of biological species that evolve into a technologically advanced civilization. We will take the optimistic assumption that given 2 or 3 billion years of evolution, this will always happen, i.e., $E=1$. Note that D and E are biological factors. $F=$ the fraction of intelligent civilizations that will be able to communicate with us. We assume that half may be aquatic civilizations that never communicate beyond their planet. Thus $F=0.5$. $L=$ the length of time that an intelligent civilization exists. This is the largest unknown in the equation although it should be remembered that some of the prior estimates may also be significantly in error. Note that F and L are sociological factors. Substituting the values listed above for the factors yields $N=(20\times0.5\times1)\times(0.2\times1)\times(0.5\times L)=L$. Thus $N=L$, where N are the number of communicative intelligent civilizations coexisting with ours in our galaxy and L is the lifetime of an intelligent civilization in years. By an intelligent civilization we mean one that has discovered and utilized thermonuclear reactions, the mechanisms that power the stars.

It is very difficult to estimate L. When we consider the danger of our own destruction by a nuclear holocaust, by changes in the planet's ecology and climatology due to human stupidity and greed, by extraterrestrial catastrophes such as Earth being struck by a huge asteroid, etc., one can reach the pessimistic conclusion that L may be as little as 100 years.

On the other hand, one can argue that if a civilization reaches a certain level of scientific sophistication, it is able to completely control its environment and hence will survive indefinitely long. For example, imagine that aging and disease were conquered so that humans were essentially a race of young immortals. With the indefinite lengthening of the active career of scientists, a way is found to more fully utilize the human brain, much of which is thought to be inactive in our intellectual processes. This results in humans becoming super geniuses. They, in turn, harness thermonuclear fusion so that the oceans of the world can provide the human race with essentially unlimited energy. This scenario can easily lead to a civilization able to fully master its environment: such a civilization could have a lifetime of millions or even billions of years. The question is, which of these extreme assumptions is closer to what actually happens?

Let us try calculating a few examples using various assumed values of L and then determining the probable distance, d, to the nearest intelligent civilization coexisting with ours, assuming that the intelligent civilizations are uniformly distributed throughout our galaxy.

Values of	Pessimistic	Intermediate	Optimistic
L (years)	10^2	10^4	10^7
N	10^2	10^4	10^7
d (light-years)	4×10^3	8×10^2	80

In the pessimistic case we are alone, and we and our nearest intelligent civilization will both self-destroy long before we have any chance of contacting them (the nearest civilization is 4,000 light-years away). In the intermediate case in which the nearest civilization is 800 light-years away, we and they would have many star systems to contact before finding one another. Would a civilization's lifetime of 10,000 years be sufficient to do this? Only in the optimistic case would we obviously make contact.

Even in the optimistic case, one must ask what form the contact would take. We know that we can communicate with radio waves over vast distances. Could we actually send a spacecraft containing humans to other star systems? The problem is the vastness of the distance to be traversed. Traveling at 100,000 mph would require almost 60,000 years for a round trip to our nearest stellar neighbor, Alpha Centauri!

The only way to reduce this enormously long travel time would be to increase one's speed close to that of the speed of light, 186,000 miles/s. Moving at such speed would also take advantage of relativistic time dilation (discussed in the chapter on relativity) which would mean that the astronauts would age more slowly than their compatriots left behind on Earth. But how can we achieve those speeds? The best chemical rockets exhaust gas at only a few kilometers per second. Using multistage chemical rockets, in order to achieve a speed of $0.99c$, where c is the speed of light, for a 100 ton payload (about twice as heavy as the Apollo spacecraft) would require a first-stage rocket of mass many orders of magnitude greater than that of the known universe.

The most efficient drive that we can conceive of would be a photon rocket engine, powered by the controlled annihilation of equal amounts of matter and antimatter. These photons are emitted out the rear of the rocket at the speed of light. It is presently beyond our scientific knowledge to build this kind of rocket.

Even if our descendants can construct such a photon engine, many problems remain before interstellar travel becomes a reality. One would have to accelerate the spacecraft relatively slowly so as not to kill its occupants. One possibility would be to accelerate at g, the acceleration due to gravity at the surface of the Earth, for half the trip, decelerate at g for the second-half of the trip to a nearby star, and repeat the process returning to Earth.

Depending upon which star was to be visited, the spacecraft would weigh at least millions of tons; all of the initial fuel would be consumed in delivering the 100-ton payload roundtrip. The trip would still take years and thus food, water, and oxygen would be problems. Finally, as the spacecraft moved at its highest velocity ($0.95c$ to $0.99c$ depending upon which nearby star is to be visited) the enormous velocity of the spaceship relative to the atoms it will encounter in outer space produces yet another threat to the crew. The front surface of the ship would be bombarded by huge fluxes of atomic constituents (protons, neutrons, etc.) similar to the intensity inside a particle accelerator on Earth. The only way to protect the crew would be to use massive shielding in the front of the ship. However this shielding increases the mass of the ship, which then requires even more fuel.

The film *Star Trek IV*, as well as the other Star Trek films and television programs, speculate that the propulsion system for interstellar travel will in fact be a photon drive powered by a matter–antimatter system.

For the reasons outlined above, scientists are skeptical of claims that we are being visited by extraterrestrials. While there have been many sightings of unidentified flying objects (UFOs), most of these sightings can be easily explained provided that investigators have sufficient information about the incident. There are some sightings that are unexplained, but those who draw the conclusion that these few sightings prove that we are being visited by extraterrestrials will find the

FIGURE 3-4. An alien spaceship landing in front of the White House would provide unambiguous proof of the existence of extraterrestrials.

scientific community skeptical. The proof will have to be much more unambiguous before scientists will accept that our planet has been singled out for attention by superintelligent creatures. Figure 3-4 suggests one way of providing unambiguous proof.

Ancient Astronauts

A variation of claims by UFO advocates that we are now being visited by extraterrestrials is the assertion that Earth was visited in the past by alien astronauts. Assertions have been made that the intelligence of our ancestors was increased by the aliens, that we are their offspring, and/or that these aliens aided in the construction of some of the wonders of the ancient world.

Probably the most-mentioned structure in this regard has been the Great Pyramid in Egypt. Numerous "experts" have claimed that this huge structure could not have been built by the Egyptians: it would have taken hundreds of thousands of workers centuries to construct it.

Modern science does not believe that one needs to invoke alien superbeings in order to explain the construction of this gigantic monument. The Great Pyramid consists of approximately 2,600,000 stones weighing 2.5 tons each. The base of the Great Pyramid was leveled by digging an approximately flat bed, filling it with water, and using the nearly perfectly flat surface of the water as a reference point from which to make the base of the Great Pyramid nearly perfectly flat.

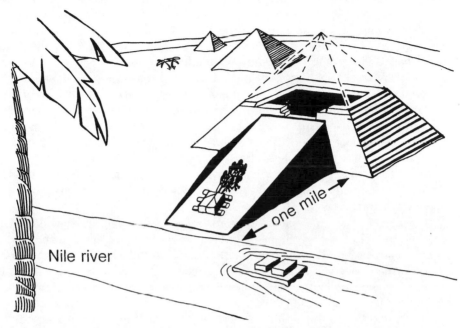

FIGURE 3-5. The construction of the Great Pyramid involved dragging building blocks up a 1-mile-long earthen rampart from the banks of the Nile River.

The building blocks were floated down the Nile on reed boats from a quarry, pulled up a mile-long earthen rampart, and then set into place, as depicted in Fig. 3-5. It is plausible that 20 workers could pull a 2.5 ton block the 1 mile from the Nile to the Great Pyramid, move it around a given rampart, and set it in place in 1 hour or so. In 1 day they could set in place 10 such blocks. Thus as few as 2,000 workers could put into place 1,000 blocks per day or 365,000 blocks per year or the entire 2.5 million blocks in 7 years. Other workers would have been needed to quarry the stones and float them down river, but it seems reasonable to assume that 20,000 workers could have built it in 20 years, as asserted by some historians. Note that one Greek historian reported seeing the remains of the earthen rampart, pictured in Fig. 3-5, centuries after the Great Pyramid was built.

Other advocates of ancient alien astronauts affecting human development have claimed that our intelligence and other characteristics were inherited from these aliens. One question often raised by these advocates is why are humans hairless? They argue that hair would have both warmed our ancestors during cold periods as well protected their relativity delicate skin against thorns, brambles, and the like. They suggest that hairlessness was inherited from alien astronauts.

There are other, less sensational, explanations for our hairlessness. One is based on a hunting theory of evolution: hairless skin would make its owner a better hunter because bare skin is a better heat exchanger and thus would allow a hairless man to literally run a fur-covered prey into the ground. After some number of miles the fur-covered animal would collapse from heat prostration and "Voila, dinner is served."

Another possible explanation of hairlessness is that our ancestors evolved along the coast of Africa during a 15-million-year drought and began using the oceans for fishing and keeping cool. Bare skin made them better swimmers. Whether either of these explanations or some other is the reason for our hairlessness is secondary to the fact that these explanations do not depend upon an extraordinary event—the intervention of extraterrestrials. The reader should be skeptical of those who propose such explanations without overwhelming proof: these claims are the bread and butter of pseudoscience and those who practice it.

Hollywood has repeatedly addressed the theme of intervention by aliens in human development. In *2001* it is the monolith that gives our ape ancestors the intelligence to use the first primitive tool—a club—which allows these ancestors to survive when faced with extinction (9 min).

The film, *Hangar 18*, deals with both UFOs and ancient astronauts. A UFO crashes; the ship is recovered intact, but the crew have been killed in an on-board accident. The scientists examining the ship decipher the aliens' language and find that they had visited Earth thousands of years ago, mating with our ancestors (66 min). Thus, they are biologically similar to us, since we are descended from them. The film does not answer the question of the origin of this alien race, nor do the scientists understand the craft's engines. No attempt is made to explain how the aliens achieved interstellar travel, nor why they have avoided contacting Earth's officials: these are the questions that are the hardest to answer for any fictional treatment of the issue.

Because of the difficulties in traveling to other stars to search for other possible civilizations, scientists think that the most promising approach is to search the heavens for radio transmissions emanating from other star systems. Attempts are underway to conduct this search for extraterrestrial intelligence, or Project SETI for short. The cost involved in such a search is extremely small (about $10 million per year) compared to the impact that such contact would have on us culturally and scientifically.

Exercises

1. *The Planets:* If the average density of the Earth is greater than the density of the rocks found at its surface, what can you conclude about the density of some parts of the Earth's interior?

2. Why is Io orange-colored?

3. In *Total Recall* the gases emitted when the alien power plant is turned on shatter the windows of nearby structures. Was the air pressure of the expanding gases larger, the same, or smaller than atmospheric pressure on the surface of the Earth?

4. *The Sun:* Why would temperature and pressure increase as one proceeds into the interior of the Sun?

5. *The Moon:* Why would the Moon have become geologically inactive billions of years ago while the Earth is still geologically active?

6. Could you lift an object more easily on the Moon or on the Earth?

7. *Meteors and Comets:* How do the destructive effects of a meteor striking the Earth vary with its mass? With its speed just before impact?

8. What kind of a defense can be built to protect Earth against being hit by an asteroid?

9. *The Galaxy and the Universe:* If we see the galaxies in all directions receding from us, does that mean that our galaxy is at the center of the universe?

10. *The Evolution of the Universe:* If gravity is the weakest of the fundamental forces, why is it the dominant force in shaping our universe?

11. Why is the Cosmological Principle important in considering the future of the universe?

12. *Intelligent Life Elsewhere:* Should we transmit signals to other worlds or only listen for their signals?

13. What kind of evidence would you find convincing that UFOs are spaceships from another star system?

14. What do you think is a reasonable value for the lifetime of an intelligent civilization?

15. *Ancient Astronauts:* What kind of evidence would you find convincing that extraterrestrials have visited Earth in the distant past?

16. Name other ancient artifacts that you have read were constructed with the aid of alien astronauts. What do you think of those claims?

Chapter Four
Electricity and Magnetism

Electrical Forces

The universe in which we live is made up of atoms, each of which contains electrical charges. One model of an atom compares it to a miniature planetary system with a dense nucleus (instead of the Sun) and electrons (instead of planets) moving about the nucleus. There are protons and neutrons in the nucleus. Each proton carries the same positive charge. Each electron carries the same negative charge, which is identical in magnitude to the positive charge on the proton. Neutrons have no electrical charge. In a **neutral atom** there are as many protons in the nucleus as there are electrons in orbit about the nucleus. Thus at a distance the atom appears to have no net electrical charge.

The reason that the protons do not pull the electrons into the nucleus is the same reason that the Moon is not pulled by gravity into the Earth; the force of attraction is completely used up putting the electrons into an orbit. Actually the planetary model of an atom oversimplifies the situation: electrons stay in a shell around the nucleus rather than in a plane of rotation the way planets move about the Sun.

Protons and neutrons are both more massive than electrons. The mass of an electron is 9.1×10^{-31} kg while the mass of a proton or a neutron is 1.67×10^{-27} kg. The value of the electric charge is expressed in units called coulombs. Protons have a charge of $+1.6 \times 10^{-19}$ C and electrons have a charge of -1.6×10^{-19} C. The forces between electric charges follow very simple rules:

Like electrical charges repel.

Unlike electrical charges attract.

Thus two positive charges will repel one another, whereas a positive and a negative charge will attract one another. It is the attractive forces between the protons in the nucleus and the electrons in orbit around it that hold atoms together. The gravitational attraction between the mass of the nucleus and the mass of the electrons is many orders of magnitude weaker than the electrical force of attraction between them. Without electrical forces atoms would not exist.

The reader may wonder why the nucleus holds together if the protons in it all repel one another. The answer is that there is a third fundamental force, the **short–range attractive nuclear force**, that holds the nucleus intact. That force will be described in the chapter on nuclear physics.

Objects are made of atoms and these atoms usually have equal numbers of electrons and protons in them. When there is an excess of either protons (because some electrons have been removed) or electrons (because some electrons have been added), the atom is electrically charged, and it is called an **ion**. An object whose atoms have an excess or deficiency of electrons is said to be electrically charged. It is important to note that when an object becomes electrically charged, no electrons are created or destroyed; they are only moved from one object to another. This is always the case and is called the **Conservation of Electric Charge Principle**:

Charge may be moved from one object to another, but no process results in a net increase or decrease in the total electric charge of the universe.

Since a charged object always has an excess or deficiency of electrons, the net charge on an object is always a multiple of the magnitude of the charge on an electron, 1.6×10^{-19} C. Thus electric charge is said to be **quantized** in multiples of the charge on the electron.

In 1785, the French physicist Charles Coulomb determined that the force between two point charges Q_1 and Q_2 is directly proportional to the product of the charges and inversely proportional to the square of the distance r, between them. As an equation, **Coulomb's Law** is written as follows:

$$F = KQ_1Q_2/r^2,$$

where the proportionality constant K in the SI system of units has the value $K = 9 \times 10^9$ N m^2/C^2. According to Coulomb's Law all charges in the universe affect all other charges, since only when r becomes infinite does the force between two charges become zero. Coulomb's Law is an inverse square law just like the Law of Universal Gravitation; if one doubles the distance between two charges, the electrical force between them decreases to one-fourth of its original value. However, gravitational forces between objects are not only weaker than electrical forces, they also are always attractive, whereas electrical forces can be either attractive or repulsive.

Materials such as metals are referred to as **conductors** of electricity because one or more of the electrons in the outer shell of their atoms are free to move from atom to atom. Other materials such as rubber, paper, and glass are referred to as electrical **insulators** because all of the electrons are tightly bound to their atoms. Still other materials can be made to behave sometimes as conductors and sometimes as insulators: these are called **semiconductors** and are widely used in the manufacture of transistors. Finally there are some materials that have zero electrical resistance: they are called **superconductors**. These materials generally become superconducting only at very low temperatures, but recent discoveries have found that there are materials that remain superconducting at much higher temperatures. This makes it economically feasible to produce very large superconducting magnets that can be used, e.g., to levitate trains. Further, power transmission lines made of these materials would have no energy losses in transmitting electric currents over long distances.

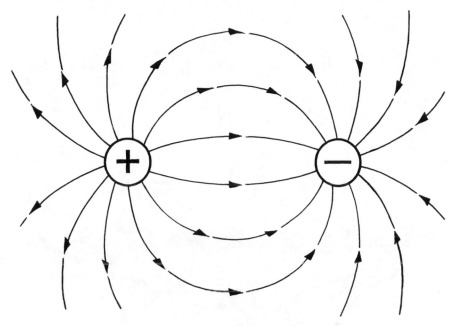

FIGURE 4-1. The electric field between a positive and negative charge.

Electrical Charges

How are objects charged? One may simply rub them together, such as rubbing a cat's fur or walking across a carpet. This is called **charging by contact**. Another way is to bring a charged object near a conducting surface. Suppose that the object is positively charged; it will attract an excess of electrons to the part of the surface nearest to the object while the remainder of the conductor will then have a deficiency of electrons. If the conductor is then separated in two parts, the part nearest to the positive charge will have a net negative charge while the other piece will have a net positive charge. Both pieces of the conductor will be said to be **charged by induction**.

The reader may wonder how an electric charge experiences the presence of another electric charge even though they are not in physical contact. The explanation is similar to that for the gravitational attraction between two masses. In the case of gravity, there is a gravitational field produced by any mass which affects space itself. In the electrical case, the first charge can be thought to create an **electric field** around itself which then interacts with the second charge. The electric field between a positive and a negative point charge is pictured in Fig. 4-1. The direction of the electric field at a point, by definition, is the direction of the force on a positive test charge placed at that point. The magnitude of the electric field is represented by the number of field lines per unit cross-sectional

FIGURE 4-2. The passenger in a car struck by a lightning bolt is shielded from the electric field of the lightning by the car's metal body.

area of space. The closer together the lines, the stronger the electric field. By these conventions, the electric field starts at positive charges and ends on negative charges, as pictured.

An important difference between electrical and gravitational forces is that we can shield ourselves from electrical forces but not from gravitational forces. A region can be shielded from electric forces by placing a conducting surface completely around it. The free electrons in the conductor will arrange themselves on its surface so as to cancel all electric field contributions inside of the conductor. If the electric field were not zero inside the conductor, the forces on the free electrons would keep moving them until there was no force on them, i.e., until the electric field became zero inside the conductor. Thus, some electronic components are encased in metal boxes and some electrical cables have a metal covering to shield them from external electrical fields. For the same reason, a car may shield its occupants from the electric field generated by a lightning bolt that strikes the car, as pictured in Fig. 4-2.

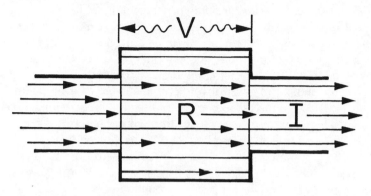

FIGURE 4-3. The current, I, flows through a resistor, R, that has a voltage difference, V, across it.

Electric Currents

Moving electric charges constitute **electric currents**. One can define an electric current as the amount of charge that passes a given cross-sectional area of a wire per second. The unit of current is the **ampere**, which is defined as the passage of 1 C of charge per second. Since the charge on an electron is -1.6×10^{-19} C, 1 C is the amount of charge on 6.25×10^{18} electrons.

There are two types of electric current: **direct current (dc)** and **alternating current (ac)**. Dc flows in just one direction in a circuit. Ac has its charges moving first in one direction and then in the opposite direction. This type of current is produced by an alternation in the direction of the voltage at the energy source. In the US, nearly all commercial ac circuits involve voltages and currents that oscillate sinusoidally at 60 cycles/s (Hz). Ac voltages and currents are used worldwide because higher voltages transmit currents with less heat loss per unit of energy transmitted over long distances. With ac voltages, transformers can be used to increase the voltage for transmission at a power generating station. Other transformers then reduce the voltage to the lower values that are used in homes and businesses.

Electric charges do not flow by themselves. A steady electric current requires a voltage source to keep the charges moving. Chemical batteries and electric generators are the most common sources of electrical potential difference. All materials (except superconductors) have a resistance to the passage of an electric current. They require an electric potential difference to overcome this resistance.

For many materials there is a simple relationship between the current, I, passing through the material of resistance, R, and the voltage difference, V, across it (see Fig. 4-3). This relationship is called **Ohm's Law**:

$$V = IR.$$

V is usually given in volts (v), I in amperes (a), and R in ohms (Ω). For a circuit of constant resistance, the current drawn is proportional to the voltage across the circuit. If the voltage is doubled, twice as much current will pass through the circuit. If you know any two of the three quantities in Ohm's Law, you can calculate the third. For example, if a light bulb has a resistance of 120 Ω, when it is connected to a 120 V circuit it will draw a current $I = 120$ V/120 $\Omega = 1$ A.

An electric shock can cause severe physical injury or even death. The severity of a shock depends upon the magnitude of the current, how long it passes through the victim's body, and where it passes through the body. A current of 10 mA (1 mA $= 10^{-3}$ A) causes muscular contraction. This prevents the victim from releasing the source of the current (e.g., a wire), and death from paralysis of the respiratory system can occur. If a current greater than 70 mA passes through the victim's torso so that a portion passes through the heart for more than a second, the heart muscles will begin to contract irregularly ("ventricular fibrillation") and death will result if the condition is not speedily ended. A much smaller electric current can cause death if it passes directly into the heart (e.g., through a catheter inserted into the heart).

The voltage difference needed to produce a given current through the human body depends upon the effective resistance of the body. This varies dramatically with the dryness or wetness of one's skin. The effective resistance between two points on a human body when the skin is dry will be 10^4–10^6 Ω, whereas when the skin is wet the resistance may be as little as 10^3 Ω. A victim with wet hands who is making good electrical contact with the ground (e.g., wet bare feet) and touches a 120 V line would have a current of 120 V/1,000 $\Omega = 120$ mA passing through his body: this could kill him.

For the above reasons it is extremely dangerous to handle any electrical device while taking a bath. One should always unplug electrical devices before looking inside them because sometimes a wire inside such a device breaks loose and touches the metal cover of the device: in that situation one could be electrocuted merely by touching the case.

Note that to receive a shock there must be a difference in electrical potential across part or all of one's body. That is why birds can perch atop high voltage wires without being killed. The potential is the same across all parts of their bodies. If a bird were to straddle a high voltage wire and another wire which is grounded, the entire voltage difference of the high voltage wire would be across the bird, and it would be electrocuted.

Electrons move relatively slowly down electric wires. This is due to the electric field that exists throughout the wire because of the applied voltage from a generator or battery. If one calculates the "drift velocity" of electrons that move along a wire, one finds that it is typically much less than 1 cm/s. Thus it would take hours for electrons at a light switch to physically move to a lamp that is turned on by the switch. Electrons have a random motion in metal wires that is much greater (they move at about 10^6 m/s), but it is random in nature and averages out to zero motion an any given direction. What makes the lamp light go on seemingly instantaneously after you turn on the wall switch is that the electric field moves down the wire between the switch and the lamp at the speed of light; this forces the electrons that are already in the lamp's filament to move and then the bulb lights up.

Electric Power

Moving charges in an electric circuit do work. The rate at which work is done, that is, the rate at which electric energy is converted into another form of energy such as heat or mechanical energy, is called **electric power**. Power was defined earlier as work/unit time. Since one form of work equals the charge $\times$ the potential difference it moves through, the power P, in watts is equal to:

$$P=(charge/time)\times potential\ difference,\ P=IV.$$

P is in watts if I is in amperes and V is in volts. If a 600 W toaster is plugged into a 120 V source it will draw a current of $P/V=600\ W/120\ V=5\ A$.

Since $V=IR$ from Ohm's Law, the expression for electric power can be rewritten $P=I(IR)=I^2R=(V/R)V=V^2/R$.

The expression to use for power depends upon which of the quantities are given and which you are trying to calculate. For example, in order to determine the current drawn by a 100 W bulb plugged into a 120 V socket you would use: $I=P/V=100\ W/120\ V=0.83\ A$.

If, on the other hand, you wanted to know the current drawn by a 1,000 W space heater that has an internal resistance of 10 Ω, you would use: $I^2=P/R$ $=1000\ W/10\ \Omega=100\ A^2$. Hence $I=10\ A$. You may wish to calculate the current that is drawn by each appliance plugged into a given circuit in your home. Usually each circuit can draw a maximum current of 15 or 20 A. If the circuit draws more than this current, a fuse will blow or a circuit breaker will open to prevent the wires in the home from becoming overheated by the current and starting a fire. Note that this heating is proportional to the square of the current flowing through the wires; doubling the current produces four times as much heat loss in the wires.

Most of the circuits in your home are parallel circuits. As pictured in Fig. 4-4, the same potential drop occurs across each device that is in parallel across a given voltage source. The total current that one draws in a parallel circuit is equal to the sum of the currents through each device. Thus you can add up the currents drawn by each appliance plugged into a given circuit in your home to determine the total current that will be drawn by that circuit if all of the devices are turned on at the same time. Here each device operates independently so that if one of them (e.g., a light bulb in a lamp) burns out, the other devices are unaffected. The other common type of electric circuit is called a **series circuit**. As illustrated in Fig. 4-5, the same current flows through each appliance connected in a series circuit. If one of the appliances connected to a series circuit burns out, the current cannot

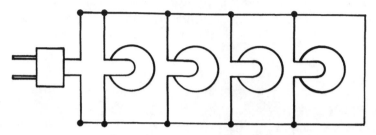

FIGURE 4-4. The same potential drop occurs across circuits in parallel, such as appliances plugged into a given circuit in a home. If the current path through one conductor is broken, the currents continue to flow through the other conductors.

continue flowing in the circuit and thus all appliances would turn off. Some strings of Christmas tree bulbs are wired in series; if one of the bulbs burns out the entire set goes dark.

In the film *The Day the Earth Stood Still*, the aliens selectively turn off electricity on the surface of the Earth. Thus, they do not cause death by stopping electric currents which operate ambulances, airplanes, etc. (61 min). This appears to be an impossibility. We do not know of any mechanism for turning off the forces between electric charges. We *can* shield something from external electric fields by enclosing it in a conductor, but in that case everything inside the container would be shielded from the external electric fields. Furthermore, if one turned off all the electric forces around the electrons (the primary charge carriers in electric circuits) the electrons circling each atom would go off into space and all of the atoms affected would collapse, instantly killing everyone on Earth. Since electric forces have an inverse square dependence on distance, it would be impossible to turn off the electric currents in one car without turning them off in an adjacent ambulance. How could all radio and television broadcasts be prevented from being transmitted in the atmosphere without also turning off the motors of airplanes passing through the atmosphere? In short, even if the aliens had the ability to turn off all electric forces, they could not have achieved the results depicted in the film.

Magnetism

In addition to the electric forces that act between charges at rest, there are another set of forces, **magnetic forces**, that act between moving charges. Both electrical and magnetic forces are different aspects of the same phenomenon, called electromagnetism.

Coulomb's Law provides us with the exact force between charges at rest, but it does not describe the forces between moving charges. In order to understand magnetic forces we will first consider "permanent magnets." A bar magnet, such

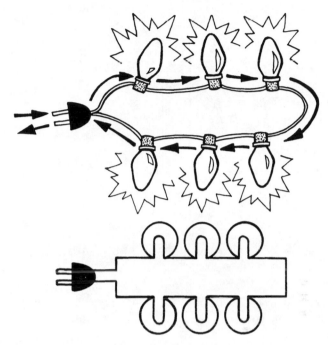

FIGURE 4-5. The same current flows through each device that is connected in series, such as a string of Christman tree lights. If the filament in any one bulb breaks, the entire string of bulbs goes dark.

as the one shown in Fig. 4-6, that is allowed to rotate will point in a north–south direction. The end of the bar magnet that points toward Earth's geographic north is called the north pole of the magnet. The end that points toward geographic south is called the south pole of the magnet.

When the north pole of one magnet is brought near the north pole of another magnet they repel, while when brought near a south pole they attract. One can generalize this by stating:

Like magnetic poles repel, unlike poles attract.

This rule is similar to that for the forces between electric charges. However, unlike electric charges which can be found singly, magnetic poles always appear in pairs. One never finds an isolated magnetic north pole or south pole. If you break a bar magnet in two, you have two complete bar magnets, each with a north and south pole. Even if you broke the magnet down into pieces the size of atoms, you would still have north and south poles. This suggests that the atoms themselves are magnets. What is the origin of their magnetism? It is the motion of the electrons in the atoms. Electrons in an atom both revolve around the nucleus and

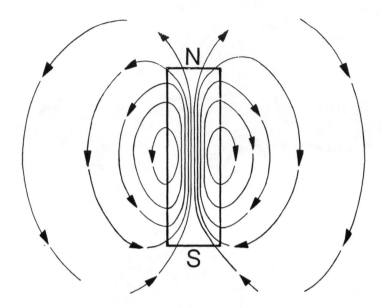

FIGURE 4-6. A bar magnet. The magnetic field lines go from the northern end of the magnet and return at the southern end.

spin on their axes. Both of these motions can contribute to the magnetism of the atom since sometimes the spin of the electrons will produce an additive effect magnetically.

For some materials, such as iron, the individual magnetism of each atom links with nearby atoms to form **magnetic domains**. These domains are microscopic in size. If these domains are randomly oriented, their effects cancel out. If the domain become aligned, as in Fig. 4-7, by bringing another magnet nearby, then the entire piece of iron becomes a new bar magnet. Dropping or heating a bar magnet will cause some of the domains to become randomly oriented, and the magnet as a whole becomes weaker.

The relationship for the force between two magnets is more complicated than Coulomb's Law for the electrical force between charges. However, we can describe what is happening conceptually by saying that the magnet sets up a **magnetic field** around itself similar to the gravitational field around any mass or the electric field around any charge. The magnetic field then interacts with any other magnet exposed to it.

In addition to spinning electrons inside atoms producing magnetic fields, an isolated moving electric charge or an electric current also produces magnetic fields. The magnetic field that surrounds a current-carrying wire can be seen by bringing a compass close to the wire. The compass will always line up with the resultant magnetic field in any region of space. By using such a compass one finds

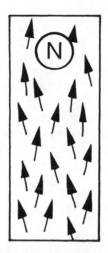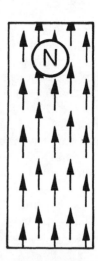

FIGURE 4-7. As the initially unmagnetized domains become progressively more aligned, they produce a progressively stronger bar magnet. The heads of the arrows point toward the north pole of the magnet.

that the magnetic fields around current-carrying wires are concentric circles about the wire. If one bends the wire into loops, one concentrates the magnetic fields inside the loops. If we then place an iron core inside the current-carrying coil, we can fabricate a very strong electromagnet. You may have seen electromagnets strong enough to lift entire cars in a junkyard.

Even stronger superconducting magnets produce enormous magnetic fields which are maintained very economically because the superconductors have no electrical resistance and, hence, a current moving through them produces no heat loss. They are powerful enough to lift entire train cars. These levitated train cars are capable of traveling at high speeds since they are not slowed by friction from the tracks. The force to lift the train cars off the tracks comes from the repulsion between the magnets in the bottom of the train and the magnetic image induced into the metal surface above which the train moves.

In addition to producing magnetic fields, moving charges also interact with magnetic fields. The force on a moving charge due to a magnetic field is unlike any of the other forces we have described so far. It acts at right angles to the velocity of the charged particle and to the magnetic field in which it is moving. If the velocity of the charged particle (or current) is reversed, the force on it is also reversed. This is the basis for electric motors.

The Earth's magnetic field has been of enormous importance to the human race. The forces exerted by the Earth's magnetic field on charged particles from outer space have prevented these particles from striking the surface of the Earth. Some of the charged particles have been trapped in orbits around the Earth. Thus, the surface of the Earth has been spared bombardment by many of the so-called cosmic rays, the nuclei of atoms stripped of their electrons. There have been times when the Earth's magnetic field was much weaker than it is today; during such periods the greater amounts of cosmic radiation reaching the Earth may have resulted in a larger number of mutations than usual.

The Earth's magnetic field, approximately one-half gauss (a unit used to specify the magnetic field) at its surface is believed to be due to electric charges moving in the liquid metal which comprises the Earth's outer core, about 1,800 miles below its surface. This liquid core, about 2,160 miles in radius, has a solid metal inner core, believed to consist mainly of iron. Changes in the currents circulating in the liquid core have produced changes in the Earth's magnetic field. The direction of the Earth's magnetic field can be determined from the way iron atoms aligned themselves in molten rocks which cooled into solids at various times in the past. The alignment of these iron atoms gives these rocks a slight permanent magnetism whose orientation is the direction in which the Earth's magnetic field was pointing when they solidified.

Computers

Another important technological application of magnetism is in the functioning of computers. Computers use a binary system for coding and manipulating information. In this binary system, all information is coded either 1 or 0 ("on" or "off"). Each statement of information is referred to as a "bit." Magnetism is also used in storing information in this binary system. Computer disks, audio and video tapes, even credit and automated teller machine cards use magnetic disks, film, or tape. Information is coded on the material, again using a binary system where every available section of material is either magnetized (a 1) or unmagnetized (a 0). The reading of the disk, film, or tape is performed by an instrument which decodes and translates the information to its original format.

Computers also use electric currents to "read" information in their memories and to perform calculations. A computer cannot be made too compact because these electrical currents passing down its wires produce heat which must be dissipated to prevent damage to the circuitry. One way to avoid this heating problem is to use superconducting wires and circuits. With the recent discovery of "high-temperature" superconductors described earlier in this chapter, the possible use of superconductors in computers will be reexamined. Earlier efforts to use superconducting thin films in computers failed because of difficulties in manufacturing films to an exact standard.

A computer has three advantages over a human in performing tasks. It is faster, it does not make errors, and it has a virtually unlimited memory. Today's fastest electronic computers can perform several billion calculations per second. The human brain can perhaps perform one or two calculations per second. To see the difference this would make in doing repetitive calculations, suppose that in the near future a supercomputer could perform 32 billion calculations/s, each of which would take 1 second for a human to perform. Then in 1 second the computer could perform as many calculations as it would take a human, working 24 hours a day, 365 days a year, 1,000 years to perform (there are about 32 million seconds in a year).

A computer's memory can be extended virtually indefinitely. By contrast the human brain is relatively limited in the amount of information it can store. One may roughly estimate the "storage capacity" of the human brain as follows. Researchers have suggested that there are about 10^{10} neurons in a typical human brain. These neurons are the active elements in the brain that perform calculations, for example. Each neuron may be connected to 1,000 or so adjacent neurons through links, called synapses. If one multiplies the number of neurons by the number of synapses one may have an estimate of the number of bits of information that a human brain can store at any given time. The calculation yields $10^{10} \times 10^3 = 10^{13}$ bits of information. This is a large number, but it can be exceeded by a computer: in principle, it could be far exceeded by a computer.

The human brain functions impressively even though it has less storage capacity than a computer, processes information much more slowly, and makes mistakes. Humans can reason, use judgment, and are creative. A computer does not exercise judgment or reason. It simply calculates according to formulas programmed into it.

A computer's operation can, at times, seem to mimic human intelligence. For example, computers have been programmed to play chess at the level of high-ranked human chess masters. But the computer plays in a different mode than humans. It will use "brute force" methods of looking at most or all possible variations for several moves into the future. In a matter of seconds or minutes it will analyze thousands or even millions of positions resulting from every legal move that can be made on the chessboard for each player. Since there are typically 30 or so such legal moves, there are $30 \times 30 = 900$ possible positions resulting from each legal move a player and her opponent can make. If one considers all possibilities just two moves ahead, the combinations grow to nearly a million positions. Human players never use the brute force method. Intuition, which is developed by past experience, reduces the possibilities a chess master examines to a few moves only. Recently, world chess champion Gary Kasparov easily beat one of the best chess-playing programs. Computer programmers confidently assert that the day will come when the world chess champion will be a computer. But when, or if, that will occur is not certain.

A recurrent theme in science fiction films is computers that are intelligent and capable of independent action. The computer HAL in *2001* and *2010* is an example of an immobile computer (as contrasted with a mobile computer—a robot—which we will consider next) which is intelligent. HAL suffers the equivalent of a human nervous breakdown in *2001* because of conflicting orders programmed into it. When it learns that the crew is about to deactivate it, it kills four of the five crewmen (82 min). Until, if ever, we have sentient (self-aware) computers it is hard to speculate on whether this behavior is reasonable or not. Certainly, protecting oneself from termination seems perfectly logical.

The film *Colossus: The Forbin Project* describes a supercomputer, placed in charge of America's missile defenses, that starts to think for itself. At one point in the film the President is told by Dr. Forbin that the executive programming unit refuses to function: the computer is acting independently (44 min). No explanation is given as to how this happened. Rather, Dr. Forbin states earlier that Colossus is working even better than anticipated. There is an unexplained increase in its speed; it performs test calculations in only 6.45 rather than 1100 cpu s (20 min). But this 200-fold increase in its speed would simply mean that the computer would continue to operate as programmed, only 200 times faster. It would not mean that Colossus could make independent judgments. Perhaps the film is implying that Colossus modified its own programs, and even circuits, in order to perform more efficiently. Even if this occurred, it would not mean that Colossus was sentient. There are computers today capable of fine-tuning their own programs. From the computer's viewpoint, joining forces with its Soviet counterpart, Guardian, does achieve its prime directive of protecting the US from attack, although not in the manner envisioned by its builders, since the two computers in effect become world-wide dictators.

There is also no explanation for the very human feelings that Colossus has for Dr. Forbin (in the books on which the film is based, Colossus refers to Dr. Forbin as "father"). Forbin tries to undermine Colossus, and as the computer states at the end of the film, no one poses a greater threat to it. Yet, Colossus assures Forbin that he will soon grow to love it (98 min). This conduct is hard to explain unless the computer has developed human emotions as well as intelligence.

In *Terminator 2: Judgment Day* we are told that the day Skynet, America's defense computer, became sentient it started World War III in an attempt to eliminate humans, who were trying to disconnect it (67 min). This film's depiction of the actions of intelligent computers is thus much gloomier than that of *Colossus*. In addition, the robots in *Terminator 2* are mobile computers, and we see from time to time the list of possible actions appearing inside the robot's "brain." How its programs manage to select the probabilities for each course of action in order to determine what to do is not indicated. Nor does the film explain the apparent emotions that the Terminator has for the boy it is protecting.

In *Blade Runner*, the moving conclusion of the film states that the androids sought answers to the same questions that humans ask—"where am I going and how much time have I got?" (107 min). This film suggests that with artificial intelligence will come emotional needs very similar to that of humans. In short, one could not develop androids that would act independently "on the job" without also acting independently "off the job"—having a personal life and needs.

Exercises

1. *Electrical Forces:* If you triple the distance between two charges, by what factor does the electric force between them change?

2. Assuming that the proton and electron in a hydrogen atom are 10^{-10} m apart, calculate the electric and gravitational forces between them.

3. Suppose that someone posted a sign on the door of a laboratory which stated, "Danger 100,000,000 Ω." Comment on the probable purpose of the sign.

4. The wattage rating on light bulbs assumes that they are plugged into a voltage source of 120 V. How many amperes flow through a 100-W bulb plugged into a 120-V source?

5. Suppose that one had three 100-W light bulbs connected in parallel to a 120-V source. How much current will the three bulbs draw? What is the resistance of each bulb? What is the value of the single resistance that would draw the same current as the three bulbs draw?

6. Why might an electrician wear gloves and rubber-soled shoes when handling live electric wires?

7. *Magnetism*: Since the Moon cooled sooner than the Earth, it is reasonable to assume that it no longer has a molten metal core. If that is the case, what conclusion would you draw about the magnetic fields around the Moon?

8. In terms of magnetic terminology, is geographic north on the Earth a magnetic north pole or a magnetic south pole?

9. How much work is done on a charged particle moving in a region of space where the magnetic field is uniform and perpendicular to the path of the particle? Describe the motion of the particle.

10. If you were to demagnetize a bar magnet, what would you have done to the domains in the magnet?

11. *Computers*: Why do you need to store computer disks away from permanent magnets?

12. Is the hard drive in a computer likely shielded by a metal casing? Why or why not?

Chapter Five
Atomic and Nuclear Physics

The Atom

As discussed in Chap. Four under Electrical Forces, the model of the atom is one of a positively charged nucleus surrounded by a cloud of electrons moving in shells about the nucleus. The diameter of an atom is typically 10^{-10} m. Often the atomic size is expressed in units called angstroms (Å), where 1 Å equals 10^{-10} m. The electrical force of attraction between the nucleus and the electrons holds atoms together. In this chapter we will examine the nucleus itself and various emissions which are given off by atoms.

In an electrically neutral atom, the number of protons in the nucleus is identical to the number of electrons moving in the surrounding cloud. When atoms come close together, the negative electrons of one atom may at times be closer to the positive nucleus of another atom, which results in a net attraction between the atoms and their formation into molecules.

The atom is mostly empty space. The nucleus and surrounding electrons occupy only about 10^{-15} of the volume of the atom. If it were not for the electrical forces of repulsion between the electrons of neighboring atoms, solid matter would be much more dense that it is.

As we know, atoms come with quite different properties. **The main characteristic that distinguishes one type of atom (or element) from another is the number of protons in the nucleus.** Hydrogen is the simplest element and consists of a single electron in orbit around a single proton. The next simplest element, helium, has two protons and two neutrons in its nucleus surrounded by two electrons. Each successive element has an additional proton in its nucleus and an additional electron moving about the nucleus. We classify atoms by their **atomic number**, which is the same as the number of protons in the nucleus. Thus, hydrogen has the atomic number 1, helium 2, and so on. Naturally occurring uranium has an atomic number 92. Artificially produced elements have even higher atomic numbers.

Electrons move about the nucleus in concentric shells at various distances from it. Each shell is limited in the number of electrons it can contain. For example, in the innermost shell there are at most two electrons, and in the second shell there are at most eight electrons.

Antimatter

In contrast to matter, which is composed of atoms with positively charged nuclei and negatively charged electrons, antimatter is composed of atoms with negatively charged nuclei and positively charged electrons, called *positrons*.

Positrons were first discovered in 1932; today, antiparticles of many types have been produced in the laboratory. Antiprotons, for example, have the same mass as protons but are negatively charged. All subatomic particles have been found to have corresponding antiparticles. Gravity does not distinguish between matter and antimatter, nor is there any way of telling whether an object is antimatter by studying the light that its atoms emit. Only by contact could we tell that an object is antimatter, for when matter and antimatter meet, all of the material is converted into radiant energy. This annihilation results in the maximum energy output per unit mass of substance since there is a 100% mass conversion into radiant energy. Nuclear fission and fusion, to be discussed later in the chapter, convert far less than 1% of matter into energy.

There cannot be substantial amounts of antimatter in our immediate vicinity, for it would annihilate as soon as it contacted matter. If, e.g., the Moon were made of antimatter, a release of energy would have resulted as soon as one of our spaceships touched it, and the spaceship would have been completely annihilated. We thus know that the Moon is not made of antimatter, nor are other planets on which we have landed vehicles. Whether there is antimatter in the composition of other stars or galaxies is not known.

Atomic Spectra

One may study the radiation emitted from atoms to seek clues to atomic structure. Each element has its own characteristic pattern of electron energy levels. When an electron moves from a higher energy level to a lower energy level, the atom emits a characteristic pattern of light frequencies, which is called its emission spectrum. Fig. 5-1 is an artist's depiction of a portion of the emission spectrum for the simplest element, hydrogen. The instrument used to view spectra is called a **spectroscope**, and is depicted in Fig. 5-2. Atoms are heated or an electric current is passed through them. The light that is then emitted by the atom is first passed through a thin slit and then focused through lenses onto a diffraction grating or prism and finally onto a screen. Each component color is focused at a different position according to its wavelength and forms an image of the slit on the screen. The different-colored images of the slit are called **spectral lines**, as depicted in Fig. 5-1.

The study of the spectra of many elements led to the conclusion that electrons in atoms could exist only in certain discrete energy levels. This contradicted classical mechanics, which permitted the electrons to orbit a nucleus in a continuum of energy states. The discrete amounts of radiant energy emitted by atoms when electrons move from higher energy states to lower energy states could only be explained by the assumption that the possible energy states of the electron were discrete in the atom. This quantization of the possible energy levels in the atom led to quantum mechanics, one of the two major developments in physics in the twentieth century (the other is relativity).

FIGURE 5-1. Part of the spectrum of hydrogen.

The explanation for these quantized energy levels was that each electron had a characteristic wavelength associated with it. The circumference of the innermost orbit of an electron in an atom is equal to one wavelength of this electron wave. The wave must close in on itself, i.e., the circumference of the orbit must be an integral multiple of this characteristic wavelength. This concept is similar to a bracelet made of equal-size loops. No matter how long the bracelet, it is always equal to some multiple of the diameter of a single loop. Since the circumferences of the electron orbits are discrete, it follows that the energy levels are also discrete, since they depend upon the radial distance of the electron from the positively charged nucleus.

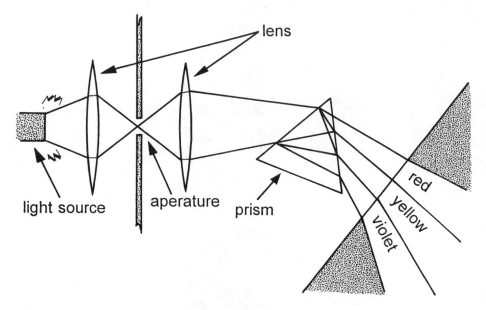

FIGURE 5-2. A schematic of a spectroscope.

This model of the atom also explains why electrons do not spiral into the nucleus, causing atoms to shrink into oblivion. If each electron orbit is described by a standing wave, the circumference of the smallest orbit can be no smaller than a single wavelength. It should be noted that in this picture of the atom, the electron is no longer thought of as a discrete particle existing at some distinct point in the atom. Its mass and charge are considered to be spread out throughout this wavelength, as if it were a cloud rather than a distinct particle. Figure 5-3 depicts the orbit of an electron in which the circumference is equal to six wavelengths. In summary, particles have wavelike attributes.

Similarly, waves have particlelike attributes. Albert Einstein received the Nobel Prize for his 1905 explanation of the photoelectric effect, the emission of electrons from the surface of a metal. This explanation involved the assumption that light, which was thought to be a wave phenomena, traveled through space and interacted with matter in discrete packets of energy called **photons**. The energy of a photon is $E=hf$, where h is Planck's constant (6.6×10^{-34} J s) and f is the frequency of the electromagnetic radiation in cycles per second or hertz.

The space ship in *Star Trek IV: The Voyage Home* used a matter–antimatter converter for its power source and also used high energy photons to rebuild a "dilithium crystal" in its power source (34 min). Theoretically, matter–antimatter annihilation is the most efficient possible energy source, but the problem not addressed by the film is how to shield the crew from the radiation produced. In other words, all of the radiant energy produced by the conversion of matter into photons would not be emitted only out of the rear of the starship but rather would

wavelength = L

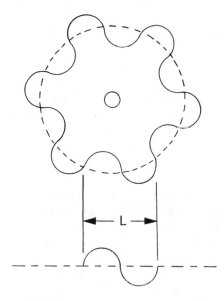

FIGURE 5-3. An electron in orbit about a nucleus can only exist in an orbit which is equal to an integral multiple of its wavelength.

pass throughout the ship, irradiating all on board. Remember that the high energy photons produced in matter–antimatter annihilation do not carry any charge. Thus, magnetic fields would be useless as shields against them.

Another term used throughout the *Star Trek* series is "photon torpedoes." These nasty weapons are pictured as pinpoints of brilliant light launched against an adversary with devastating effects. Based upon the formula for the energy of a single photon, it would require an immense number of them to have significant destructive power.

Finally, *Star Trek IV* (as well as all the other *Star Trek* film and television productions) uses a transporter device which breaks down individuals into electrical impulses to transport them elsewhere (73 min). Consider the transporter's operation for a moment: each of the 10^{27}–10^{28} atoms that comprise the human body must be broken down and converted into electrical impulses and reassembled at its destination. But, what reassembles these 10^{27} or more distinct atoms which have arrived at the target as pulses of light? There is no transporter to receive them. Furthermore, there is no known physical principle which can act to convert matter completely into electromagnetic energy pulses except a matter–antimatter reaction. And that reaction would be explosive, killing not only those transported but destroying the ship as well. There is no plausible way of doing this.

It is interesting to note that the idea of using a transporter device instead of landing the *Enterprise* was reportedly introduced in the original *Star Trek* television series because it lowered production costs. One did not need to build a model of the spaceship to be placed on a costly set depicting an alien planet. And, of course, it opened many plot possibilities. But, it is still impossible.

The film, *The Fly*, also dealt with the transportation of a man between teleporting and receiving units (34 min). A computer disassembled and reassembled the individual (unfortunately confusing his atoms with that of a fly). Most of the objections against the *Star Trek* teleport are again applicable—it just could not happen.

In *The Adventures of Buckaroo Banzai*, the hero has developed a jet car that passes through an entire mountain at 700 mph (7 min). The invention that made this feat possible was called an oscillation overthruster. It so weakened the electrical bonds between the nucleus and the electrons that the car could pass through the mountain. The overthruster supposedly produced colliding beams of electrons and positrons, which by their annihilation reduced the range of the electromagnetic force to less than the size of the nucleus, allowing the car to pass through essentially empty space. It had to complete the journey quickly, before the atoms returned to their original state. In the film Buckaroo correctly states that atoms are mostly empty space, and that nuclei typically take up only one-quadrillionth the volume of an atom (31 min).

There are several problems with the oscillation overthruster idea. First, the energy released in the matter–antimatter collisions of the colliding beams would be lethal to the driver of the car. Second, the atoms would fly apart; as soon as the electrons no longer experienced the electrical attraction of the positive nucleus, they would not continue in circular orbits. Instead they would fly off, producing more radiation and leaving a large hole in the mountain. Finally, the striking of the car by an alien entity from the eighth dimension as it passed through the mountain is pure fantasy.

In *Terminator 2: Judgment Day*, the evil terminator can assume the shape of anything it comes in contact with and can pass through, e.g., iron bars (56 min). In order to do this, the terminator's atoms would have to recognize the atoms of the bar on an atomic level and change themselves to replicate them. Thus, each change undergone by this terminator involves the relocation of each electron and proton in its entire structure, an impossibility.

The Nucleus

In 1896, the German physicist Wilhelm Roentgen (1845–1923) discovered an unknown radiation, which he called x rays, that were produced by an electron beam striking a glass surface. These x rays penetrated through some materials better than through others, and could produce x-ray photographs of the human skeleton.

Today we know that x rays are high-frequency electromagnetic waves, which are usually emitted by the excitation of the innermost orbital electrons of atoms

The wavelengths of this radiation were typically on the order of 1 Å, whereas visible light has a wavelength of 4,000–7,000 Å. The production of x rays today is done in tubes which accelerate electrons through voltages of 30,000–150,000 V before they strike a glass or metal target, from which the x rays are emitted. In addition to their medical uses, x rays have been used to explore the very atomic structure of matter.

Shortly after Roentgen announced his discovery of x rays, other researchers discovered that uranium naturally emitted radiation, as did other elements such as polonium and radium. In these cases, the emitted radiation was the result not of changes in the energy states of the electrons of the atom but of changes occurring within the central atomic core, the nucleus. These rays were emitted as a result of the disintegration of the atomic nucleus, which we call **radioactivity**.

Radioactive elements emit several types of radiation. Alpha particles are helium nuclei, which have a positive electrical charge. Beta rays are fast-moving electrons with a negative electrical charge. Gamma rays are short wavelength electromagnetic radiation (shorter, even, than x rays) which have no charge at all. If one places the three kinds of radiation in a perpendicular magnetic field, alpha particles are deflected in one direction, beta rays in another, and gamma rays pass through undeflected.

The nucleus is composed of **nucleons**, which are protons and neutrons. How, then, is an electron ejected from the nucleus (beta particle emission)? A neutron within the nucleus is converted into a positive proton and a negative electron. The proton remains within the nucleus while the electron is ejected from it. The radius of a nucleus varies from about 10^{-15} m for hydrogen to about 6×10^{-15} m for uranium. The shapes of nuclei are generally spherical. There are energy levels within the nucleus, just as there are energy levels for the electrons in orbit about the nucleus. Changes of energy states within the nucleus lead to the emission of gamma rays.

The protons and neutrons in the nucleus tend to lump together into alpha particles (two protons and two neutrons). According to quantum mechanics, there is some probability that an alpha particle can escape the attractive nuclear force that binds it to the nucleus. Once it is outside, but immediately adjacent to the nucleus, its two positively charged protons are strongly repelled away from the remaining protons in the nucleus, and **alpha particle emission** has occurred.

In addition to alpha, beta, and gamma rays, a large number of other particles have been detected as originating from the nucleus. Perhaps the most important of these is the neutron. The number of neutrons in the nucleus of a given element may vary somewhat. Atoms that have identical numbers of protons but different numbers of neutrons are called **isotopes** of a given element. Scientists use a shorthand notation for a given element which has a unique number of protons in its nucleus. For example, U stands for uranium. The superscript to the right of the symbol then gives the atomic mass number, i.e., the number of protons and neutrons in the nucleus. The subscript to the left of the symbol gives the number of protons in the nucleus. Thus, $_{92}U^{235}$ represents an isotope of uranium that has

92 protons and $235 - 92 = 143$ neutrons in its nucleus. Sometimes the subscript is omitted, since all uranium nuclei have 92 protons. Thus, one may also identify the isotope by writing U^{235}.

Since like electrical charges repel, there must be another force to hold the positively charged protons together in the nucleus. This is called the nuclear force, and is attractive. This force is identical between two protons, two neutrons, or a neutron and a proton. It extends only for very short distances, on the order of 10^{-15} m. Nuclear forces are much stronger than electrical forces. Since they only interact between essentially adjacent nucleons, while the repulsive electric forces act between all the protons in the nucleus, it is clear that larger nuclei containing more protons are less strongly held together than small nuclei, and may be subject to spontaneously breaking apart. Relative stability is achieved in the heavier elements by having more neutrons than protons. The greater number of neutrons is required to compensate for the repulsion between distant protons.

A nuclear force distinct from the strong nuclear interaction is responsible for beta emission. This weaker nuclear force is called the weak interaction and largely affects lighter particles like electrons and even lighter particles called neutrinos. It should be noted that we have now discussed all four of the fundamental forces found in nature: gravity, electricity and magnetism, the strong nuclear interaction, and the weak interaction.

Nuclei having many protons, such as uranium, are unstable. It is not possible to predict when a given atom of a heavy element will disintegrate, a process called **radioactive decay**. However, we can predict the **half-life** of a radioactive isotope. It is the time needed for half of a given number of atoms of that isotope to decay. Each half-life results in half of the remaining undecayed nuclei undergoing disintegration. In two half-lives, only one-quarter of the original nuclei will remain undecayed. The half-life of some elements is as short as a millionth of a second, while U^{238} has a half-life of 4.5 billion years.

One can use the half-life of certain isotopes to date artifacts. In living organisms, there is a fixed ratio of the radioactive carbon 14 to the more abundant carbon 12. When a plant or animal dies, the percentage of carbon 14 decreases with a half-life of about 5,700 years. The radioactivity of living things therefore gradually decreases at a steady rate after they die. We can measure the radioactivity of plants or animals today and compare this with the radioactivity of ancient organic matter. If, for example, we find that a piece of wood has one-half as much radioactivity as an equal quantity of wood extracted from a living tree, then the wood (or an object made from it such as an axe handle) must have come from a tree that was cut down about 5,700 years ago. Using carbon dating we can accurately establish dates up to about 40,000–50,000 years ago.

One can use uranium dating to establish the age of rocks, since U^{238} decays through several stages to become lead^{206}, whereas the common lead isotope has an atomic mass of 208. Knowing the half-life of U^{238} and U^{235}, scientists have established the ages of rocks on the Earth and on the Moon to be as old as 4.2 billion years.

Effects of Radiation

What is the effect of radiation on human beings? Radioactivity has been around since the Earth was formed. It warms the interior of the Earth and makes it molten. Table 5.1 provides some average radiation exposure levels from the most common sources of radiation.

TABLE 5.1. Exposure levels from common radiation sources.

Radiation source	Exposure level
Diagnostic x rays	0.04–0.10 REMs/yr (estimates vary)
Natural radioactivity	0.02–0.20 REMs/yr
Cosmic rays	0.035 REMs/yr (sea level)
	0.060 REMs/yr (5,000 ft elevation)

The REM, (Rad Equivalent Man), is the unit of radiation that has already taken account of the fact that different kinds of radiation have different effects on humans. For example, a given amount of energy absorbed from neutrons will do more biological damage than the same amount of energy absorbed from x rays. Note that for most persons the majority of radiation is received from natural sources.

Natural radioactivity sources include radon in the air, and radiation from the soil, building materials, food, and water. The amount of radiation received from medical x rays varies substantially depending upon the number and type of x rays received. Note also that the radiation from cosmic rays increase as one goes to higher elevations, since there is a thinner atmosphere above the individual to absorb the radiation from outer space. Thus, a couple of round-trip flights across the U. S. at 30,000 ft or higher may expose the traveler to 0.01 REMs, about as much radiation as from a dental x ray.

It is prudent to reduce exposure to unnecessary radiation whenever possible. It is not certain what level of radiation is the maximum to which the general population could be safely exposed. Some scientists have suggested that, on average, whole-body exposure for the general public to radiation should not exceed 0.5 REMs/yr. It is generally assumed that it is safer to receive a given quantity of radiation in small dosages over a long period of time than in one single dose. Presumably, the human body has evolved in a manner so that it can tolerate small dosages of radiation. Table 5.2 presents the effects of a single short exposure to the specified level of radiation.

TABLE 5.2.

Exposure level (REMS)	Effects
1	No detectable body change.
10	Blood change barely detectable.
400	Vomiting, fatigue, damage to bone marrow, infection, bleeding, temporary sterility. Approximately 50% die in 60 days.
600–800	Approximately 80%–100% die within 60 days. Effects include damage to the central nervous system

While exposure to radiation can cause cancer, it can also be used to treat it. Rapidly growing cancer cells are especially susceptible to destruction by radiation. Nonetheless, surrounding normal cells are inevitably killed as well, and for this reason cancer patients receiving radiation therapy often suffer side-effects characteristic of radiation sickness. To reduce the destruction of normal cells, narrow cross beams of gamma rays or x rays can be used, if the tumor is localized. The beam is always directed at the tumor, but the source of the beam is rotated so that it passes through various parts of the body to keep the damage to the normal cells as low as possible. In other cases, a tiny radioactive source may be inserted directly inside the tumor, which will eventually kill the majority of the cells. An example of this technique is the use of radioactive iodine to treat cancer of the thyroid.

Other applications of radiation include sterilizing bandages, surgical equipment, and even packaged food, since both bacteria and viruses can be killed by large doses of radiation. Finally, radiation is used in radioactive tracers, which can be used to follow bodily functions in detail.

Radiation Detectors

There are several types of radiation detectors in use. The simplest is the radiation badge, which is a piece of unexposed film; exposure to radiation will cause a clouding of the film.

The Geiger counter consists of a central wire in a hollow metal cylinder filled with gas. An electrical voltage is applied between the cylinder and the wire. If radiation enters the tube and ionizes an atom in the gas, the freed electron moves toward the positively charged central wire. As it collides with other atoms along its path, more electrons are knocked out, resulting in a large number of electrons moving toward the wire. This creates a short pulse of electric current and is registered on the instrument along with a clicking sound.

A more sophisticated instrument, a scintillation counter, uses materials that emit light when charged particles or gamma rays strike them. These light emissions are

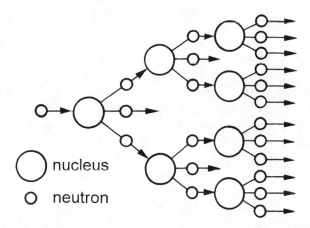

FIGURE 5-4. A chain reaction in which each disintegrating nucleus releases three neutrons, two of which strike other nuclei, causing their disintegration. The large circles represent nuclei and the small circles represent neutrons.

converted into electric signals in photomultiplier tubes. A scintillation counter is more sensitive to gamma rays than a Geiger counter and can also measure the energy of the charged particle or gamma ray absorbed by the detector.

Nuclear Fission and Fusion

When heavy nuclei disintegrate, energy may be released. The question arose: could a way be found that would accelerate the disintegration of energy-releasing nuclei such as U^{235}? In 1939, two German scientists found that they could split the uranium nucleus by bombarding it with neutrons. **Nuclear fission** is the name given to the process of breaking a large nucleus into two or more smaller nuclei. If the combined masses of the smaller nuclei are less than that of the parent nucleus, the mass loss is converted into radiant energy.

In U^{235}, each disintegration includes the release of two or three additional neutrons, which may, in turn, strike other U^{235} nuclei, causing them to disintegrate with the release of yet more neutrons and more energy. Figure 5-4 illustrates this process, called a **chain reaction**. A chain reaction requires a sufficient number of U^{235} atoms (an amount called the **critical mass**) so that the neutrons emitted by one disintegrating nucleus will strike other U^{235} nuclei rather than leaving the sample without undergoing another collision. The chain reaction occurs in a small fraction of a second, and the tremendous amount of energy released by the large numbers of disintegrating nuclei results in an explosion. This is how an atomic bomb explodes.

The amount of U^{235} needed to build an atomic bomb is quite small! The uranium used in the Hiroshima bomb was about the size of a baseball. One simply needs to have subcritical masses of U^{235} brought together suddenly in order to trigger an atomic bomb. The problem in building the first atomic bomb was that uranium in nature is 99.3% U^{238} and only 0.7% U^{235}. The U^{238} would not undergo fission but could absorb the neutrons, thus preventing a chain reaction from occurring. Thus, the U^{235} had to be separated from the U^{238}. Uranium isotope separation today is accomplished using a gas centrifuge. The heavier U^{238} moves to the outside of a drum rotated at very high speed, and the lighter U^{235} is extracted from the center of the rotating drum.

Another way of obtaining fissionable material is by transforming the abundant U^{238} into plutonium^{239}, which is fissionable in the same manner as U^{235}. Nuclear reactors can be used to convert U^{238} into plutonium^{239}. Such reactors are called **breeder reactors**, because they effectively produce more fuel than they consume.

A nuclear reactor consists of rods of nuclear material, such as U^{235}, which are placed in graphite moderators that function to slow down the fast neutrons emitted by the disintegrating nuclei (slower neutrons are more likely to interact with other U^{235} nuclei). In addition, the reactor contains rods of cadmium, or boron steel, which can be moved into or out of the graphite in order to increase or decrease the absorption of neutrons, and thereby slow down or speed up the nuclear reaction. In short, a reactor is a way of producing a controlled chain reaction. Reactors are heavily shielded to protect workers from radiation emitted by the radioactive elements in the reactor. The heat emitted by the reactor is used to produce steam, which drives an electric generator in the same manner as the heat coming from the combustion of coal or oil in more conventional electrical generators.

Scientists have several concerns about the operation of nuclear reactors. The first is a safety concern: are the reactors designed and operated in such a way to avoid the possibility of another Chernobyl disaster? In that tragedy, the containment vessel around the reactor ruptured and radioactive materials were hurled into the sky. This heavily contaminated the surrounding area and was even carried by winds into foreign countries. A properly constructed reactor and containment vessel would have avoided the Chernobyl disaster.

A second concern of scientists is the disposal of radioactive waste. Some of the radioactive waste products have very long half-lives. One must store them in a manner to ensure that they will not reach water supplies or otherwise contaminate the environment. Simply storing them in metal drums, for example, would not be sufficient if the half-life of the radioactive material is much longer than the time it will take for the metal drums to deteriorate so that their contents are released to the environment.

A third concern is keeping track of the nuclear material entering and leaving a reactor. The amount of fissionable material needed to build an atomic bomb is quite small. Terrorists may try to obtain enough fissionable material to build a suitcase-sized atomic bomb.

In addition to nuclear fission as a proces of converting mass to energy, there is **nuclear fusion**. In this multistep process, four hydrogen nuclei (ie., protons) are combined into one helium nucleus. This process is accompanied by the conversion of mass to energy. This is the process which powers our Sun and other stars. It is difficult to initiate, since positively charged protons repel one another. They do not normally come close enough together for the attractive nuclear force to bind them into a helium nucleus. Only if the protons have very high kinetic energy can they overcome the powerful Coulomb repulsion of other protons in order to form helium.

To date, scientists have been able to bring the protons together only by raising their temperature to millions of degrees and thereby sufficiently increasing their kinetic energy. One way to achieve these high temperatures is to detonate an atomic bomb. Thus, the thermonuclear, or hydrogen, bomb has an atomic bomb as its trigger. When the atomic bomb is detonated, the temperatures inside the fireball ignite the thermonuclear reaction of the hydrogen to achieve the thermonuclear explosion.

Unlike the atomic bomb, there is no limit to the size or explosive power of a hydrogen bomb, because its nuclear fuel will not detonate until ignited by the explosion of an atomic bomb. The atomic bombs which destroyed Hiroshima and Nagasaki each had an explosive power of about 20,000 tons of TNT. Each bomb consisted of subcritical masses of uranium which were brought together by conventional explosives, resulting in a chain reaction. Practically, one cannot simultaneously bring together more than a few subcritical masses of uranium. Thus, the explosive power of an atomic bomb cannot exceed more than that of about 100,000 tons of TNT. By contrast, the largest hydrogen bomb exploded to date had the explosive power equivalent to 68,000,000 tons of TNT!

In principle, one might build a nuclear fusion reactor which would use the oceans of the world as a virtually inexhaustible source of thermonuclear power, since water contains hydrogen as well as oxygen. In addition, there would not be radioactive waste products from a thermonuclear reactor. However, it is very difficult to achieve and sustain the enormously high temperatures needed for the thermonuclear reaction. Scientists are using high-powered pulsed lasers in order to momentarily achieve these high temperatures. A further problem is how to contain the fast-moving protons at these temperatures. Since all known materials melt at a temperature of a few thousand degrees, scientists are experimenting with magnetic fields which are configured so as to confine the fast-moving protons until they interact with one another. Commercial fusion reactors are not likely to be available for several decades.

Forbidden Planet displayed the power output of 9,200 thermonuclear reactors on gauges on the walls of the alien Krell laboratory (58 min). While the maximum power output of a single reactor was not stated, Dr. Morbius referred to them as the power of an exploding planetary system, and at the end of the film their explosion does destroy the planet, Altair 4. As described in Chap. 2, each reactor is seen delivering about 10^{12} W with only about 16 of the dials lit. Does this mean that all 30 or more of the gauges would register if all of the reactors were working at full capacity? With 31 gauges lit the reactors would be supplying 10^{31} W.

Dividing this by the 10^4 reactors would imply that each could produce 10^{27} W full scale. This seems hard to believe, since it would involve the conversion of about 10^{10} kg (or 10 million tons) of matter into pure energy per second in each reactor! This is more than twice the amount of matter converted into energy per second in our Sun. Perhaps the last several gauges were, thus, not meant to be used.

In the film *Aliens*, the thermonuclear reactor plant on a distant world explodes like a hydrogen bomb (123 min). This catastrophe is hard to understand. Presumably it was caused by the troops in the film blowing holes in the plant's cooling system while firing at the alien creatures. But if the cooling system were damaged, the plant would presumably melt down, not explode. In the reactor core, the components would melt long before the temperatures rose to the millions of degrees needed for the hydrogen to fuse into helium on a large scale. The melting of the reactor core would then dissipate the raw materials for the thermonuclear reaction, not bring them together.

In addition, the plant's computer warns of the impending disaster long before its occurrence (109 min). The plant should have been built so that in an emergency, the equipment stopped working rather than continued to deliver hydrogen to the thermonuclear core. Thus, if there was advanced warning of a breakdown, it should have shut down the plant. This shutdown might have led to a meltdown of the core, not to a thermonuclear explosion. The origin of the electrical discharges that are seen inside the plant just before the explosion is not clear.

Elementary Particles

Particle accelerators have probed the interior of nuclei and nucleons with high-speed projectiles. The cyclotron was developed in 1930 by E. O. Lawrence (1901–1958). It uses a magnetic field to maintain charged ions (usually protons) in nearly circular paths. Their path is separated by two gaps. During each circle they pass twice into the gap and a voltage is applied to accelerate them. This increases their speed and also the radius of curvature of their path, as indicated in Fig. 5-5. After many revolutions, the protons have acquired a very high kinetic energy and have reached the outer edge of the cyclotron. They then either strike a target placed inside the cyclotron, or are directed by other magnets to an external target.

A synchrotron is a sophisticated cyclotron which deals with the fact that the mass of the accelerated particles will increase as their speed approaches the speed of light (see the chapter on relativity). These accelerators can provide protons with energies of 10^{12} eV. An electron volt equals 1.6×10^{-19} J and is the electrical energy that an electron would gain by being moved through a voltage difference of 1 V. In the mid-1930s, elementary particles consisted of neutrons, protons, electrons, positrons, photons, and neutrinos for a total of six elementary particles. Elementary particle physics, as it exists today, began in 1935 when the Japanese physicist, Hideki Yukawa (1907–1981), predicted the existence of a new particle that was involved in creating the strong nuclear force. He called it a **meson**. Since then, about 200 other elementary particles have been detected in nature.

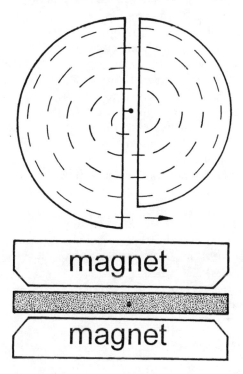

FIGURE 5-5. A cyclotron. The top figure depicts the instrument from above while the bottom figure is a side view of the machine. The cyclotron itself is sandwiched between two electromagnets, which produce a magnetic field perpendicular to the path taken by the charged particles inside the cyclotron.

The film, *Fantastic Voyage*, involved the miniaturization of matter so that a submarine and its occupants could be inserted into the blood stream of a dying scientist in a desperate effort to save his life (25 min). The film states that this process could only be applied to atoms but not nuclear materials (19 min)! Since every atom has nuclear material at its center, this provides a logical contradiction in the plot. More importantly, such a process violates conservation of energy. Suppose that one of the occupants of the submarine had a mass of 100 kg when normal-sized. When that mass is down to a fraction of a gram, where has all the rest of the mass gone? And what has happened to all of the atoms in his body, and to their constituents such as protons, neutrons, electrons, and other elementary particles. Finally, where does the energy come from to cause them to return to their normal mass after 1 hour? It simply cannot happen.

In the *Outer Limits* episode, "The Production and Decay of Strange Particles," we have a strange radiation suddenly building up at a nuclear power facility (1 min). A new heavy element has been placed in a cyclotron and results in a mysterious radiation that somehow interacts with isotopes already in the nuclear "furnace." The director of the nuclear facility claims that the radiation has also torn open a hole into another dimension, resulting in a seething glaring light that kills the reactor workers and replaces them with mysterious entities that appear to consist of glaring blue light presumably composed of these strange particles. These beings draw their power from the mysterious radiation, and they act in concert to widen the hole. This story is rather hard to swallow. The heavy element mentioned in the story had an atomic mass of 256, slightly more than uranium, which certainly does not behave as the element in the film when placed in a reactor. Also, the story claims that the nuclear reaction is "going critical" and the material will presumably explode. This is impossible in a reactor in which the control rods are in place. Finally, the solution to the problem was to detonate a thermonuclear device (47 min). But this would not, as the film claims, somehow reverse time, and thus the process caused by the strange radiation. Rather, it would simply cause an explosion.

Strange particles are so named because they are always produced in pairs and their lifetimes are longer than first expected. These facts about strange particles hardly imply that they can think and act cooperatively, or replace workers at nuclear plants!

Exercises

1. *The Atom*: How many electrons are in orbit about a uranium nucleus?

2. *Antimatter:* What would happen if the Earth collided with a 1,000 kg meteor made of antimatter?

3. *Atomic Spectra*: Calculate the energy in one photon of visible light of frequency 5×10^{15} Hz.

4. *The Nucleus*: Shortly after Roentgen discovered x rays, a number of state legislatures banned the production and sale of x-ray opera glasses. What prompted these actions and why were they unnecessary?

5. If a certain radioactive substance has a half-life of 1 minute, how many atoms of the substance remain undecayed after 4 minutes, if there are initially 1,024 such atoms?

6. If the amount of radioactive carbon 14 in a wooden artifact is only one-eighth of a new piece of the same wood, how old is the artifact?

7. *Effects of Radiation*: One article suggested that the effects of one REM of radiation on one's life span was the same as smoking 143 cigarettes. What does this imply about smoking?

8. *Nuclear Fission and Fusion*: How might you bring together subcritical masses to manufacture an atomic bomb?

9. Can a nuclear reactor undergo an atomic-bomblike explosion?

10. What limits the effective size of a hydrogen bomb?

11. Using nonrelativistic physics, calculate the speed of an electron moving

with a kinetic energy of 10^{12} eV. If you accept the fact that nothing can move faster than the speed of light, 3×10^8 m/s, what conclusion do you draw from this calculation?

12. What happens when an electron and a positron meet?

Chapter Six
Relativity and Time

In 1905, Albert Einstein proposed the special theory of relativity. The word "special" is used, because this theory deals only with reference frames that are not accelerating. This theory was based upon two postulates. Einstein's first postulate stated:

All laws of nature are the same in all frames of

reference moving at constant speed in a straight line.

Such a frame of reference is called an **inertial frame of reference** because Newton's First Law, which defines inertia, properly describes motion in such a frame of reference. Thus, an object at rest or moving at constant speed in a straight line with respect to an inertial frame of reference will continue in that state of rest or uniform motion unless a resultant force acts on it. Such a frame of reference can be an observer, a train, a car, some point on the Earth, the Sun, a spaceship, etc. Actually, the Earth is not an inertial frame of reference since it rotates on its axis once every 24 hours, moves in a circle about the Sun once a year, and as part of the solar system, moves in a circle in our galaxy about once every 225 million years. However, for many situations one can neglect these motions of the Earth and instead assume it to be an inertial frame of reference. For the same reason, a tree or a sign post fixed to the Earth is often assumed to be at rest.

Einstein's first postulate simply states that no physical experiment can be performed to determine whether the observer is at rest or moving in a straight line at constant speed. For example, on a jet airplane flying in a straight line at 500 mph, tea pours exactly as it does if the plane is sitting on the runway. In fact, if an airplane's occupants cannot see their motion relative to the ground, there is no way for them to determine how fast they are moving on the basis of any experiment that they can perform inside the airplane cabin. They must make measurements of physical parameters outside of the airplane to determine its speed.

One should note that while the laws of physics are the same in all inertial frames of reference, different observers will see the same event differently, as predicted by the laws of physics. For example, imagine that the driver inside a car moving at constant speed drops a ball out of the window of the car. Consider the ball's motion as seen from two different frames of reference, that of the car and that of the road on which the car is traveling. From the point of view of the car, the ball will be accelerated vertically downward from the hand of the driver who drops it. This acceleration is caused by the attraction of the Earth on the ball. Relative to the driver, the ball does not have any forward velocity and thus falls vertically downward. On the other hand, an observer at rest on the road will see the ball move in an arclike path to the ground since the ball has a forward velocity (that of the car

reference frame = car

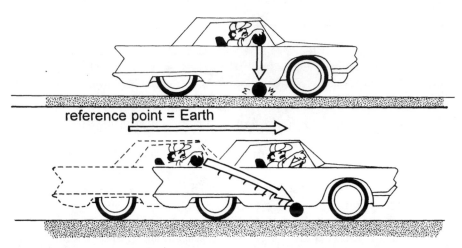

reference point = Earth

FIGURE 6-1. The motion of a ball dropped from a moving car as seen from the car's reference frame (top of figure) and from the side of the road (bottom of figure).

and its driver) with respect to the road and, in addition, is being accelerated vertically downward by gravity. Figure 6-1 depicts the ball's path in the two different inertial frames of reference. There is no contradiction because the two observers see the ball moving differently. In each case, Newton's laws of motion properly describe the way the observer sees the ball move.

Einstein had a second postulate, which seemed more radical than his first postulate. The second postulate stated:

Light propagates through empty space with a definite

speed C independent of the motion of the source of the

light or of the observer.

This second postulate is at variance with classical mechanics. For example, suppose the driver of a car, moving at 93,000 miles/s relative to the Earth, measured the speed of the light leaving the car's headlights to be 186,000 miles/s. Classical mechanics would predict that an observer standing by the roadside must measure the speed of light of the car's headlights to be $93,000 + 186,000 = 279,000$ miles/s. Yet, an extensive set of experiments conducted by A. A. Michelson and E. W. Morley found that the speed of light in a vacuum was always exactly 186,000 miles/s, independent of the motion of the source of the light or the observer. This result was one of the great puzzles of physics at the end of the nineteenth century. Various farfetched theories were devised to gain agreement between experiment and classical mechanics, including one which conjured up something called "ether," whose sole property was to yield agreement with classical mechanics!

How could the speed of light be measured to be the same in all inertial systems? Since speed=distance traveled/time elapsed, Einstein concluded that both the units of distance and time varied with the relative speed of the inertial system in order to make the speed of light come out to be exactly the same value. This challenged the basic concepts of space and time which had been considered self-evident for centuries.

Time Dilation

Time is not absolute but is relative to the motion between the observer and the object being watched. Suppose that a clock measures a time interval t_0 when it is at rest with respect to the observer. When the clock moves at a speed v with respect to the observer, it will appear to tick more slowly as seen by the observer. The observer will measure a time t on an identical clock at rest with respect to the observer while the moving clock is registering a time period t_0. The two times are related as follows:

$$t=t_0/(1-v^2/c^2)^{1/2}.$$

Thus, if a clock ticks a hundred times when it is moving at $v=0.5c$, i.e., at half the speed of light with respect to an observer, an identical clock in the observer's hand (which measures the time t) will tick 115 times while the moving clock ticks only 100 times, i.e., $t=1.15t_0$. Nothing is wrong with the moving clock; rather it is ticking at a different rate dictated by the different space-time continuum in which it exists. This phenomena is called **time dilation** and has been confirmed experimentally many times by measuring the apparent increase in the lifetime of fast-moving radioactive particles as their speed is increased relative to the laboratory in which the time t is measured. The relationship between t and t_0, the lifetime measured for the radioactive particles when they are at rest with respect to the laboratory, always follows Einstein's time-dilation formula exactly.

Figure 6-2 presents another way of seeing that the time will be measured as passing more slowly on a clock in a reference frame moving with respect to the observer. Suppose that the clock measures the time it takes for a pulse of light to go from a light source to a mirror and then to a receiver. Then, on board the moving vehicle the light travels a nearly vertical path up to the mirror, and back down to the receiver. Its total path length is about $2D$, where D is the distance between the light source and the mirror. However, the observer on Earth sees the capsule which holds the clock moving to the right and thus, that observer measures the light as traveling a greater distance from the light source to the receiver. Consequently, the Earth-bound observer will measure a longer period of time for

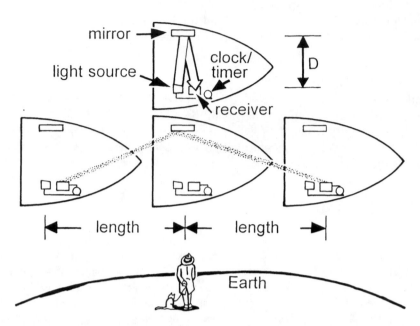

FIGURE 6-2. The distance light travels from a source to a mirror to a receiver as seen from the inside of a spaceship holding the apparatus (top of figure) and from the Earth (bottom of figure).

the light to travel a greater distance, i.e., it takes a longer time between "clicks" on the moving clock, as measured from the Earth-bound observer's time frame.

A human being's biological clock would also "tick" appreciably slower if that individual were traveling at speeds close to the speed of light, since a biological clock would undergo the same dilation phenomena that a mechanical clock undergoes. Figure 6-3 depicts a famous illustration of the effects of time dilation. It is the classic case of twins, one of whom departs on a round-trip space voyage at speeds approaching that of light, while his brother stays behind on Earth. Upon reunion, the astronaut twin is younger than his Earth-bound brother. How much younger depends upon the length of the trip, and the velocities and accelerations experienced by the spacecraft in its round-trip voyage to another star.

Time dilation could have very dramatic effects if some, as yet unknown, power source were discovered which could maintain a constant acceleration for a spaceship equal to that of gravity on the surface of the Earth, namely 9.8 m/s². Astronauts would feel very comfortable aboard such a ship, since they would experience an acceleration identical to that of Earth's gravity. Suppose that an astronaut accelerates at this rate for 10 years, then reverses the direction of the ship's rockets and decelerates at the same rate for another 10 years, lands on the planet of a distant star and then repeats the process to return to Earth. The astronaut will have aged 40 years; Earth will have aged thousands of years! In fact, observers on

FIGURE 6-3. A twin leaves for a round trip to a nearby star: he returns to find that his twin brother is now older than himself.

Earth will see that the astronaut has traveled a distance it takes light thousands of years to travel. Thus, time dilation makes it possible, in principle, for the human race to travel throughout our galaxy if only we had a method of providing the tremendous amount of energy needed to sustain accelerations in order to reach speeds very close to that of light. We would also have to devise a method of protecting the astronauts from bombardment by subatomic particles that would be striking them at a relative speed very close to that of light.

The reader may wonder why the twin brother who leaves on the spaceship does not see his twin on Earth age more slowly, i.e., why is there no symmetry between what the space-bound brother and the Earth-bound brother predict? There is no paradox at all. The consequences of the special theory of relativity can be applied only by observers in inertial reference frames. The Earth can be considered to be such a frame for the purpose of analyzing this problem, whereas the spacecraft clearly is not. It is accelerating or decelerating throughout its voyage. During these accelerations, the spacecraft's predictions based on special relativity are not valid since it is not moving at constant speed in a straight line. Only the prediction of the Earth-bound brother is correct.

Time Travel

Time dilation might lead the reader to conclude that time travel is, at least in principle, possible. This is correct but it is a very limited, one-way time travel. In principle, if scientists discovered new power sources and methods of shielding our astronauts, we could send them into the future in the sense that they would return to Earth thousands of years from now. However, they could never travel backwards in time to tell us what the future holds. Our last direct physical contact with them would be when they blasted off for their destination. One cannot devise a time machine that moves its occupants backwards in time.

Note also that the lifetime of the time travelers remain unchanged in terms of their life experience. In the case of the twin paradox described above, the brother who returns 40 years later has experienced only 40 years of life, not the thousands of years that have elapsed on Earth. In short, the time travelers do not achieve personal immortality in the sense of experiencing thousands or even millions of years of life. They experience, on average, the same number of years as their compatriots left behind on Earth.

There is another type of "time machine" which can look backwards in time. That time machine is the sky in the sense that events that took place long ago are just now reaching our eyes when we look up at the sky at night. The light from our nearest neighboring star, Alpha Centauri, left it 4.3 years ago, and when we look at it in the night sky we are seeing it as it was 4.3 years ago. Using our most powerful telescopes we can see light which left distant galaxies about 10 billion years ago, i.e., we are looking at events that occurred 10 billion years ago when the universe was much younger. This is one of the reasons that astronomers try to see such distant objects, because it permits them to look back to earlier times in the evolution of our universe. This kind of "looking back" in time is not the same as the "traveling back" in time of science fiction.

Many science fiction films discuss time travel. For example, *Star Trek IV: The Voyage Home* sends the crew of the *Enterprise* back into the 1980's in order to find whales to take into the future with them to save Earth from destruction. They achieve time travel in the film by accelerating their spaceship close to the Sun, which increases the speed of the craft and sends it back to Earth hundreds of years in the past (29 min). The film confuses two quite different phenomena, speed and time. The Sun's gravitational field could be used to speed up the spaceship, but all that would result would be a much faster moving spaceship, not a spaceship existing centuries earlier. Furthermore, going around the Sun in the opposite direction would not reverse time and send the spaceship into the future.

This idea, that motion in one direction is equivalent to time going forward, and motion in the opposite direction is equivalent to time going backward, is also used in *Superman: The Movie*. Near the end of the film, Superman decides to change the course of history in order to save Lois Lane from being killed (118 min). He zooms around the Earth in a direction opposite to its rotation, thereby reversing time (according to the film). As in *Star Trek IV*, the two physical phenomena–the direction of rotation and the passage of time–have nothing to do with one another.

In *Terminator 2: Judgment Day* we are not given any mechanism for the time machine which sends two robots into the past, one seeking to kill a future leader of the human race and the other to save him. The machine can only send living flesh into the past (5 min) and not inanimate objects such as weapons. This seems rather puzzling since one of the robots is entirely made of liquid metal, hardly living flesh. More importantly, the viewer is left with the logical paradox of any time travel into the past. If the time traveler dramatically changes the past, how can the future exist from which the time traveler came? For example, suppose that a time traveler kills his grandfather, who therefore does not have a child, who in turn does not produce the time traveler. Then how could the time traveler exist in the first place to commit the assassination? Some films have suggested that the time traveler would be prevented in some manner from changing the past in order to avoid the above paradox. In *Terminator 2* the statement is made that the future can be changed (79 min), but that would imply an infinity of parallel universes in which time travelers would jump back and forth creating different futures.

Time travel into the past also violates conservation of energy, because additional matter is suddenly added to the past from the future. The past gains energy in the form of matter, while the future loses that energy. In short, based upon the current knowledge of science and logic, time travel into the past appears to be impossible.

Perhaps the most famous time-travel film of all is *The Time Machine*, based upon the H. G. Wells novel of the same name. In the film, the time traveler can move into the future by pushing forward the control on his machine and can move into the past by pulling back on the same control mechanism (11 min). Unfortunately, traveling in time is not the same as driving a car: there is no reverse!

Length Contraction

Einstein also concluded that the unit of length would appear to decrease as it moves relative to an observer. If a ruler had a length L_0 when at rest with respect to an observer, when it moves at a relative velocity v, its measured length while moving becomes

$$L = L_0(1 - v^2/c^2)^{1/2}.$$

This change of length occurs only in the direction of motion of the object. At 87% of the speed of light an object would appear to be contracted to half its original length. If $v = c$, i.e., the object moves at the speed of light, its length would be zero. For ordinary speeds of, say, 60 mph $= 0.017$ miles/s the fraction v^2/c^2 $= 8 \times 10^{-15}$, which results in too small a change in length to measure.

Note that if v is zero, then the Einstein equations reduce to those of Newton, namely, $t = t_0$ and $L = L_0$. Thus, at ordinary, everyday speeds, time clocks and meter sticks are not seen to change appreciably, which is why the conclusions of relativity seem so foreign to us. We move too slowly to experience their effects in our daily lives.

In *Star Trek IV: The Voyage Home*, as the spaceship enters warp speed (traveling faster than the speed of light) in preparation for the attempt at time travel, the ship seems to lengthen rather than contract (30 min). As seen from a viewer at rest with respect to the Sun, which is the position from which the film views the ship, it would start to contract and then vanish as it reached the speed of light.

Increase of Mass with Speed

Einstein stated that when work is done on an object to increase its speed, its mass increases as well. Thus, an impressed force produces less and less acceleration as speed increases. The relationship between mass and speed is given by

$$m = m_0/(1 - v^2/c^2)^{1/2}.$$

In this formula, m_0 is the **rest mass** of the object, i.e., the mass an observer would measure for an object at rest with respect to the observer, and m is the mass the observer measures for the object moving at speed v with respect to the observer's reference frame. At 50% of the speed of light, $m = 1.15m_0$. If $v = c$, $m = m_0/o$ = infinity! Thus, no material object having a nonzero rest mass can move at the speed of light. This means that the universe consists of two kinds of objects: light, which can only move at the speed of light, and objects which have a nonzero rest mass, which can never attain the speed of light.

Atomic particles such as electrons and protons have been accelerated to speeds in excess of $0.99c$. Their masses have increased thousands of times, as evidenced by the path they follow when a beam of particles is directed into a deflecting magnetic field. The more massive particles do not bend as easily as they would if their mass had not increased so dramatically. They follow paths predicted exactly by the relativistic equation for the increase of mass with speed. Once again, at normal speeds of say 60 mph, this minuscule increase in mass is not measurable.

Mass-Energy Transformation

Probably the most important technological implication of special relativity is the finding by Einstein that mass is simply another form of energy. This implies that mass can be converted into energy. In typical chemical reactions, the mass is reduced by about one part in 10^9. In nuclear fission the mass conversion is one part in a thousand. Thus, the masses of the fission fragments are about a thousandth less than the mass of the initial uranium atoms and the mass difference is converted into pure energy. Einstein's formula for the relationship between mass and energy is perhaps the most famous in twentieth-century science:

$$E = mc^2,$$

where E is the energy (in joules in the SI system), m is the mass in kilograms, and c is the speed of light which equals 3×10^8 m/s. Thus, if 1 kg of matter (weighing about 2.2 lb on the surface of the Earth) is converted into pure energy, it yields 9×10^{16} J of energy, enough to keep a 100-W bulb burning for about 30 million years.

This mass to energy conversion is the basis for nuclear power and for both atomic and hydrogen bombs. The mass conversion in the Sun is due to the thermonuclear fusion of hydrogen into helium: it is the source of the sunlight so necessary to Earth. From this formula we can see that if a particle that had a nonzero rest mass were accelerated to the speed of light, its mass would be infinite and, hence, it would take an infinite amount of energy to accelerate the particle to the speed of light.

General Theory of Relativity

Einstein also considered the case in which the reference frame was being accelerated. The general theory of relativity deals with the more general case of accelerating frames of reference. Einstein postulated the **principle of equivalence**, which states that observations made in an accelerated reference frame are indistinguishable from observations made in a Newtonian gravitational field.

Consider being in a sealed spaceship far away from the gravitational pull of any star or planet. In such a spaceship moving at constant speed in a straight line, the occupants of the ship would float freely. They would feel no gravitational pull and there would be no up or down. An astronaut would weigh nothing inside such a ship. Suppose that the motors of the spaceship were then turned on so that its acceleration was exactly equal to that of gravity at the surface of the Earth. If the occupants could not see outside of the ship they would be unable to determine whether the ship was accelerating or was at rest on the surface of the Earth under the influence of the Earth's gravitational field.

No experiment that the astronauts could perform inside the spaceship could distinguish between the two possibilities: (A) that the spaceship was accelerating at g, the acceleration due to gravity at the surface of the Earth; or (B) that the spaceship was at rest on the surface of the Earth. These experiments include electromagnetic phenomena as well as mechanics. Thus, a ball dropped in the accelerating spaceship would fall to the floor in the same manner as if the ship were stationary on Earth.

Similarly, the light ray entering the ship horizontally would be deflected toward the rear of the ship, as depicted in Fig. 6-4. Einstein stated that light bends because it travels in a space-time geometry that is bent. The presence of mass causes this bending of space time, or it can be caused by an acceleration, as in the case of the rocket ship in Fig. 6-4, which pictures light deflecting due to the acceleration of the moving spaceship or the pull of gravity in a stationary spaceship (inset). The mass of the Earth is too small to appreciably bend the surrounding space-time, which is practically flat, so any such bending of light in our immediate environment is difficult to detect. However, close to bodies of much larger mass, such as the Sun, the bending of light is large enough to detect.

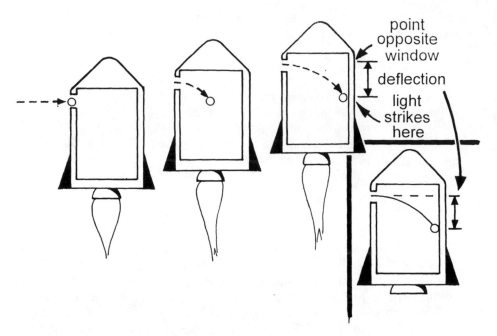

FIGURE 6-4. A beam of light passing through the window of an accelerating spaceship is deflected downward with respect to a point opposite to the window in the same manner that a gravitational field would deflect the beam downward if the spaceship were at rest on the Earth, as depicted in the figure inset.

Einstein predicted that star-light passing close to our Sun would be deflected by an amount large enough to measure. Stars are not visible when the Sun is in the sky. During an eclipse of the Sun, however, stars are visible and hence the deflection of star-light can be measured. The measurements taken during the total eclipse of 1919 confirmed Einstein's hypothesis. To see that we could not ordinarily measure the bending of light in the Earth's gravitational field, consider that the beam would fall a vertical distance of 4.9 m in 1 s (the same as a football would), but will travel a horizontal distance of 3×10^8 m in that same 1 s. This deflection would not be noticeable that far from the beginning point of the light.

Gravity and Time

Einstein's general theory of relativity predicts that gravity causes time to slow down. The stronger the gravitational field the more time slows down. This slowing down applies to all clocks—whether mechanical, chemical, or biological. Since atoms emit light at specific frequencies characteristic of the energy levels of the electrons within the atom, every atom can be considered to be a clock. A slowing down of atomic vibrations indicates the slowing down of atomic clocks. Since red light is at the low frequency end of the visible spectrum, a lowering of frequency shifts the color toward the red. This effect is called the **gravitational redshift**. Scientists have measured the slowing down of time between the top and the bottom floors of buildings since the gravitational field is stronger at the lowest floors, closest to the Earth.

This would have dramatic effects for a massive star which began collapsing. As the star proceeds toward becoming a black hole, a clock placed near the star will run more and more slowly. Time is predicted to stop completely at the center of the black hole.

Finally, general relativity calls for a new geometry of space and time. In essence, gravity is a manifestation of the geometry of space-time. The presence of mass results in the curvature, or warping, of space-time. By the same token, a curvature of space-time reveals itself as mass. Figure 3-3 in Chap. 3 presents a simplified analogy to the four-dimensional bumps and depressions in space-time. We can imagine that Fig. 3-3 was produced by a heavy mass resting on a sheet with lines drawn on it. The heavy mass causes the indentation in the lined sheet depicted in the figure. Far away from the heavy mass, the sheet resumes its normal, flat, two-dimensional form.

In the film, *The Black Hole*, the graphics used to represent the effect of matter on space-time are very similar to Fig. 3-3 (1 min). In addition, the film depicts a trip through a black hole, inside of which various incidents coexist which occurred at different times, perhaps to suggest to the viewer that time is standing still (86 min). On the other hand, the spaceship passes through the black hole in a finite time and comes out of it unharmed. That is fantasy; the ship and its crew would have been crushed to atomic size.

Exercises

1. Why are events that are simultaneous in one frame of reference not simultaneous in a frame that is moving relative to the first frame of reference?

2. *Time Dilation*: Suppose that an elementary particle has an average lifetime of 10^{-6} s when at rest with respect to an observer. What will the observer measure its lifetime to be when it is moving at $0.87c$ with respect to the observer?

3. Suppose that a clock records time passing half as fast in another frame of reference moving at speed v with respect to the clock. What is the speed of this frame of reference relative to the clock?

4. Imagine that the speed of light was infinite. What would happen to the relativistic prediction of time dilation?

5. *Time Travel*: If we can look backward in time through a telescope, why is it impossible to affect the past?

6. *Length Contraction*: A meter stick moving at 0.90 of the speed of light with respect to the observer would appear to be how long?

7. *Increase of Mass with Speed*: When v is much less than c, what should the relativistic formula for kinetic energy become?

8. *Mass Energy Equivalence*: How much energy is released if 10 kg of matter annihilate 10 kg of antimatter?

9. *General Theory of Relativity*: Why must the Sun be eclipsed when measuring the deflection of light from nearby stars?

10. A spaceship can create artificial gravity by turning on the engines to accelerate the spaceship. This will use up the spaceship's fuel. Is there a way to accelerate part or all of the spaceship, and thereby provide an artificial gravity, without using fuel continuously?

11. *Gravity and Time*: Would a clock flying at 30,000 feet above the Earth in a jet plane speed up or slow down relative to an identical clock on the surface of the Earth, and why?

Chapter Seven
The States of Matter

In Chap. 5, we noted that all matter is composed of atoms and that the number of protons in the nucleus determines a given element. Atoms can be arranged into various states of matter. In this chapter we will describe the three most common states of matter—the solid, liquid, and gaseous states.

The Solid State

In the solid state, atoms are rigidly held together by the electrical forces between them. Many solids are composed of **crystals**, in which atoms arrange themselves close together in a regular array. Each atom vibrates about its own fixed position in a lattice but is unable to move to other sites in the lattice. Figure 7-1 depicts the sodium chloride (table salt) crystal structure. The larger spheres represent chlorine atoms and the smaller spheres represent sodium atoms. The bars joining them represent the electrical forces holding the crystal together.

There are many combinations of atoms found in nature. Some of these combinations are impossible to identify by casual inspection, and extensive laboratory tests may be required to determine which element or compound is present. This fact is overlooked in the *Star Trek* episode, "*Arena*", in which Captain Kirk is battling an alien to the death on a small planet which has been prepared by other aliens as a battlefield. Kirk is told that there are sufficient resources on the planet to create lethal weapons (22 min). He discovers substances which he identifies to be potassium nitrate, sulphur, and coal, and mixes them together to make gunpowder. Using diamonds as projectiles and fashioning a primitive gun barrel from a hollow tree trunk, he downs his opponent. But how did Kirk identify each of these minerals? Further, how did Mr. Spock, watching the battle on the Enterprise view screen, also identify potassium nitrate? Several minerals have a very similar appearance. Finally, how did Kirk know the relative proportions of each constituent to make gunpowder?

Density

The masses of the atoms and the spacing between them determines the density of the substance. **Density** is defined to be mass per unit volume. Mass may be given in grams and volume in cubic centimeters: then density is in g/cm^3. Or, mass may be given in kilograms and volume in cubic meters: then density is in kg/m^3. Water has a density of 1 g/cm^3 at 4 °C. Table 7.1 gives the densities of selected substances:

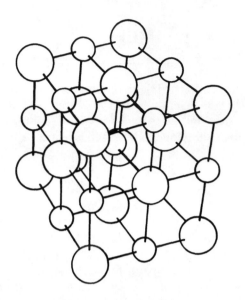

FIGURE 7-1. The crystal structure of sodium chloride. The larger spheres represent chlorine atoms.

TABLE 7.1.

Substance	g/cm^3	kg/m^3
Solids		
Aluminum	2.7	2,700
Brass	8.6	8,600
Copper	8.9	8,900
Gold	19.3	19,300
Iron	7.8	7,800
Lead	11.3	11,300
Uranium	19.1	19,100
Liquids		
Water (4 °C)	1.0	1,000
Ethyl alcohol	0.8	800
Mercury	13.6	13,600

One can use the definition of density to determine the mass of an object. For example, what is the mass of a cube of lead 2 m on a side? The mass=density×volume; thus the mass=$(11.3 \times 10^3$ kg/m$^3)$ $\times (2 \times 2 \times 2 m^3) = 90.4 \times 10^3$ kg. A related concept, weight density, is the weight per unit volume of a substance.

Size, Mass, and Strength

The dependence of mass on volume has important implications for the size and function of objects, both living and inanimate. Imagine a cube 1 m on a side: it has a volume of 1 m^3 and a cross-sectional area of 1 m^2. Now consider what happens if one increases the length of each side of the cube by a factor of 2. The cross-sectional area becomes $2 \times 2 = 4$ m^2. The volume becomes $2 \times 2 \times 2 = 8$ m^3, i.e., its volume increases by a factor of 8. Since the weight of an object is equal to its density times its volume times the acceleration due to gravity at the surface of the Earth, the weight of any uniform object will also increase by a factor of 8 if one doubles its linear dimensions.

The ability of solid objects, such as steel beams and wooden boards, to support weights placed on them is proportional to the cross-sectional area of the beam or board. If one doubles the cross-sectional area, one doubles the weight which the beam can support. Thus, a beam that is twice as wide and twice as thick as another beam can support $2 \times 2 = 4$ times the weight of the smaller beam.

This explains one of the factors in the size and strength of certain living creatures. Since muscle and bone strength are proportional to their cross-sectional area, they increase as the square of the linear dimensions of the creature. However, the weight of a creature increases as the cube of its linear dimensions. Thus, large creatures have disproportionately thicker legs than smaller creatures. The thick legs on an elephant are needed to support its great weight.

The many science fiction films that have dealt with gigantism in insects or other creatures have overlooked the above facts. For example, the film *Them!*, which depicts the discovery of giant 8-ft-long ants, argues that they would be very strong because a 1-in.-long ant can lift 20 times its own weight (50 min). Let us assume the size of one of the giant ants to be 100 in. If its strength increased in the same manner as a mammal with a skeleton (ants do not have a skeleton), this increased strength would be $100^2 = 10,000$ times the strength of a 1-in.-long ant. The giant should thus be able to lift a weight of $10,000 \times 20 = 200,000$ 1-in.-long ants. But the giant would weigh $100^3 = 1,000,000$ times as much as a 1-in. ant. The giant ant could not even lift itself! For real ants without a skeleton, the situation would be even worse.

The Liquid State

Matter can exist in configurations other than those of the solid state. Atoms and molecules in the liquid state are not restricted to vibrating about fixed points. They are able to flow, i.e., to slide over one another. **Liquids** maintain a fixed volume, but assume the shape of their containers and exert forces against the container walls.

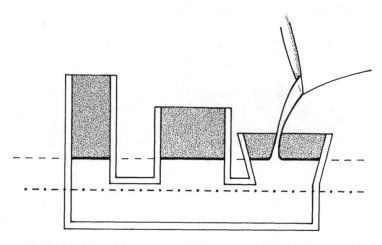

FIGURE 7-2. The level of liquid is the same in the three different shaped interconnected vessels, as is the pressure at a given depth (represented by a dashed line) below the surface of the liquid.

Pressure

In order to discuss the forces exerted in and by a liquid, it is necessary to introduce the concept of **pressure**, which is force per unit area.

$$P = F/A,$$

where P is in Newtons per square meter if F is in Newtons and A is in square meters. P is in pounds per square inch, if F is in pounds and A is in square inches.

Pressure is not the same as force. For example, an elephant weighing 10,000 lb exerts this force over its four feet with a cross-sectional area of perhaps 200 in.^{2}. This means that the pressure exerted by the elephant is 10,000 lb/200 in.^{2} = 50 lb/in.^{2}. Consider the pressure exerted by a 120 lb woman wearing very small high-heeled shoes. Her 120 lb is distributed over perhaps 2 in.^{2}, yielding a pressure of 120 lb/2 in.^{2} = 60 lb/in.^{2}. This is greater than the elephant's pressure!

As you descend below the surface of a liquid, the pressure increases linearly with your depth below the surface. This can be expressed as a formula:

$$P = \rho g h,$$

where P is the liquid pressure in N/m^2 (or lb/in.^{2}), ρ is the density of the liquid, in kg/m^3, g is the acceleration due to gravity = 9.8 m./s^2, and h is the depth, in m, below the surface of the liquid.

The pressure of a liquid is the same for a given depth below the surface of the liquid, regardless of the shape of the containing vessel. Figure 7-2 illustrates this for a variety of container shapes; the pressure 1 cm below the surface of the liquid is the same for each of the three vessels.

Liquid pressure is exerted equally in all directions. It is always directed perpendicular to the walls of any container holding a liquid. If there is a hole in the wall of the container, the liquid spurts out at right angles to that wall.

As one descends below the surface of the ocean, one needs to increase the pressure inside a diver's suit in order to decrease the pressure difference across the fabric of the diving suit. For example, at 1,000 m below the surface of the ocean, the pressure is $1{,}000 \text{ kg/m}^3 \times 9.8 \text{ m/s}^2 \times 1{,}000 \text{ m} = 9.8 \times 10^6 \text{ N/m}^2$, or about 98 times the atmospheric pressure on the Earth's surface. Thus, the air pressure inside the suit would have to be increased to 98 times atmospheric to prevent the ocean's pressure from crushing the diver.

In The Abyss, the undersea station is located 1700 feet (about 600 m) below the surface of the ocean (6 min). Since objects are lowered into the water directly from the station, the internal air pressure in the station must be equal to the pressure of 600 m of water, i.e., over 60 times the air pressure on the surface of the Earth. If the air pressure were lower inside the station, the pressure of the sea water would force water into the station through the opening into which the divers and the undersea craft enter. Anyone entering the station from the surface must spend 8 h in a special chamber in which the air pressure is increased from 1 to 60 times atmospheric.

The film shows a diver going much deeper, down to well over 17,000 ft (over 6,000 m), to deactivate an atomic warhead. It is unlikely that the human body could survive such pressure increases. In the film (108 min) the descent is made in only 30 min. How is the diver's pressure inside the suit increased from about 60 times atmospheric to over 600 times atmospheric in so short a time? What produces the heightened pressure inside the suit? The diver is breathing a liquid/air mixture, so he could not have admitted external sea water to increase the pressure, because he could not have breathed the sea water.

At the end of The Abyss, the undersea station is raised by aliens the 1700 ft to the surface in a matter of minutes (131 min). A scientist in the film correctly states that this rapid rise to the surface (without decompressing), and the resulting decrease in the pressure to which their bodies are subjected (from 60 atmospheres to 1 atmosphere), should have killed all of the occupants of the station. No explanation is given for why they are unaffected by this dramatic change in pressure.

Buoyancy

Buoyancy is the apparent reduction in the weight of objects submerged in a liquid. The origin of the buoyant force is the increase in liquid pressure as one increases the depth below the surface of the liquid. Imagine a cylinder submerged in a liquid as depicted in Fig. 7-3. The arrows represent the forces exerted on the cylinder. The forces pushing against the sides of the cylinder cancel. The force pushing upwards against the bottom is larger than the force pushing downward

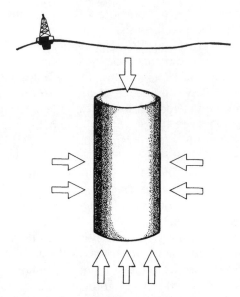

FIGURE 7-3. The forces acting upon a cylinder submerged in a liquid. The upward forces on its bottom are greater than the downward forces on its top.

against the top of the cylinder, since the pressure at the bottom of the cylinder is larger than the pressure at the top of the cylinder. This net upward force on the submerged object is called the **buoyant force**.

.Thus a boulder "weighs less" when submerged because any scale it is placed on will need to provide only some of the force needed to cancel the force of gravity: the remaining upward push is provided by the buoyant force on the rock. When the rock is lifted completely out of the liquid, it will weigh its normal amount. If the weight of a submerged object is greater than the buoyant force on it, the object will sink. If the weight is exactly equal to the buoyant force on the completely submerged object, it will remain at whatever level it is located. If the buoyant force of the completely submerged object is greater than its weight, it will float with only part of the object submerged.

An object displaces a volume of liquid equal to the volume of the object that is submerged in the liquid. The weight of that volume of liquid is the weight that is displaced by the object. If the object were not there, the volume of liquid which it displaces would be in equilibrium with its surroundings; hence, the weight of that liquid would have to be exactly counterbalanced by the buoyant force of the surrounding liquid. The atoms of the liquid that are exerting forces upon the given volume of liquid react identically, whether they are pushing against liquid atoms or against the atoms of a solid that displaces the liquid. Consequently, they continue to exert a force on the solid exactly equal to the weight of the liquid displaced by the solid. This fact is summarized in **Archimedes' Principle**:

An immersed object is buoyed up by a force equal to the weight of the fluid it displaces.

A **fluid** is a liquid or a gas, since both flow. Thus, the concept of buoyant force also applies to gases.

Submarines maneuver at a given level below the surface of the ocean by adjusting the average density of their craft through adding or expelling sea water in ballast tanks. If the craft is slightly less dense than water it will rise, if more dense it will sink.

Pascal's Principle

An important fact about liquids and gases is that a change in pressure at one part of a fluid will be exactly transferred to other parts of that enclosed fluid. Thus, if the pressure of a town's water supply is increased at the pumping station, the pressure everywhere in the pipes connected to the pumping station will be increased by the same amount. This is called **Pascal's Principle**:

The pressure applied at any point to a confined fluid is transmitted undiminished throughout the fluid.

There are many useful applications of Pascal's Principle. Figure 7-4 depicts a hydraulic lift, in which a force F_i is applied to a small piston of cross-sectional area A_i, producing a pressure P_i. This pressure is transmitted throughout the liquid to the output of the lift. Thus, $P_0 = P_i = F_0/A_0 = F_i/A_i$, and $F_0 = (A_0/A_i) F_i$. If A_0 were 100 times larger than A_i, F_0 would then be 100 times larger than F_i. In that example, if F_i was 50 lb, F_0 would be 5,000 lb, enough force to lift a car. The hydraulic press does not violate the conservation of energy principle, because the increase in force is compensated for by a decrease in distance moved. In the above example, if the input piston moved 100 in., the output piston would move only 1 in. Since work=force×distance moved, the input work=the output work, in the absence of friction.

Gases

Atmospheric Pressure

The air above the surface of the Earth exerts a pressure upon it for the same reason that water in a lake exerts a pressure on the bottom of the lake. At sea level, the pressure of this column of air is about 1.01×10^5 N/m² (or 14.7 lb/in.² in the British system of units). This is defined to be one **atmosphere's** pressure. (A related unit often used on weather maps is the **bar**, which is defined to be 1.00×10^5 N/m².) Atmospheric pressure varies with one's position above sea level. The higher the position, the lower its atmospheric pressure. Since the density

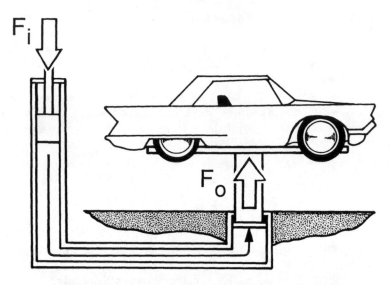

FIGURE 7-4. A car is raised by a hydraulic lift using Pascal's Principle. The force, F_i, applied to the left piston is multiplied by A_0/A_i to produce the larger output force, F_0.

of dry air near the surface of the Earth is about 1.29 kg/m^3, one can calculate the decrease in air pressure for variations in height near the Earth's surface. For example, at 1 km above sea level atmospheric pressure has decreased by about 12%. Approximately 50% of the atmosphere is below 5.6 km. The density of the atmosphere decreases rapidly above that point. About 99% of the atmosphere lies less than 30 km above sea level. This variation in atmospheric pressure with height above sea level is the reason that commercial airliners must be pressurized. They often fly at heights of 10 km or more and passengers would faint from the low air pressure and, hence, greatly reduced concentration of oxygen, if the cabins were not pressurized.

   Barometers are instruments used to measure atmospheric pressure. Figure 7-5 illustrates a simple mercury barometer. A glass tube of length greater than 76 cm and closed at one end is filled with mercury and tipped upside down in a bowl of mercury. The mercury in the tube runs out of the submerged open bottom until its height is about 76 cm above the mercury level in the bowl. The empty space at the top of the glass tube is essentially a vacuum. The barometer functions because the weight of the mercury in the tube exerts the same pressure as the outside atmosphere. If the atmospheric pressure increases, then the mercury column is pushed higher than 76 cm. These barometers are normally constructed using mercury rather than water, since mercury is 13.6 times denser than water and thus, weighs 13.6 times as much as a given column of water. A water barometer would have to be 13.6 times as high as a mercury barometer, or about 10.3 m tall to function. This is an impractical height for a laboratory instrument.

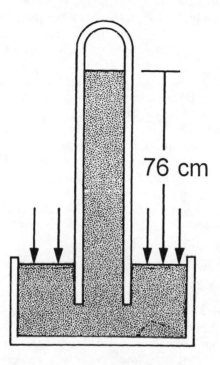

76 cm

FIGURE 7-5. A simple mercury barometer. Atmospheric pressure pushes down on the mercury in the bowl and is measured by the height of the column of mercury it supports.

The functioning of a barometer is similar to drinking a liquid through a straw. When you suck on the straw, you reduce the air pressure inside the straw that is placed in the drink. Atmospheric pressure then pushes the liquid up into the lower pressure region in the straw. Thus, the liquid is not actually sucked up; it is rather pushed up by the pressure of the atmosphere.

For a similar reason, there is a theoretical limit of 10.3 m on the height water can be lifted by a vacuum pump. The old-fashioned farm-type vacuum pump operates by producing a partial vacuum in a pipe that extends downward into the water below. The atmospheric pressure on the surface of the water simply pushes it up the pipe into the region of reduced pressure. Even with a perfect vacuum, you cannot raise water to a greater height than 10.3 m with a vacuum pump. Even Superman could not suck water through a straw higher than 10.3 m.

There are many gauges that measure pressure with respect to normal atmospheric pressure. This pressure measurement is called **gauge pressure**. Thus, to get the absolute pressure, one must add atmospheric pressure to the gauge pressure. For example, a tire gauge reads only gauge pressures. Thus, if a tire gauge registers 30 lb/in.^{2}, this is the pressure of the tire above atmospheric. The absolute pressure is $30 + 14.7 = 44.7$ lb/in.^{2}.

Blood pressure is often measured using a mercury-filled instrument. This gauge is attached to a closed air-filled jacket that is wrapped around the patient's arm at the approximate height of the heart. The air pressure in the cuff is then increased sufficiently high to prevent any blood from passing through the main artery in the arm below the cuff. The physician or nurse listens with a stethoscope at a point along this artery in the arm below the cuff. The air pressure in the cuff is then reduced until the blood begins to flow below the cuff when the heart is contracting. The listener then hears a tapping sound in the stethoscope, because the velocity of the blood flow is both high and turbulent. At this point the pressure in the cuff is equal to the **systolic blood pressure**.

As the pressure in the cuff is further reduced, blood flows past the cuff throughout the heart cycle, including when the heart is resting between contractions. The listener hears the tapping sound disappear when the blood can pass below the cuff at even the lowest pressure of the heart cycle. This pressure is called the **diastolic blood pressure**. Normal systolic pressure is about 120 mm of mercury and normal diastolic pressure is about 80 mm of mercury.

The Gaseous State

Just as water and other liquids exert buoyant forces on objects immersed in them, Archimedes' principle also applies to an object in air. Archimedes' principle explains why hot-air or helium-filled balloons can carry passengers into the air. Helium is often used in such balloons because it is much lighter than air so that the combined weight of helium, balloon, passengers, and gondola can be less than the weight of the air it displaces. However, there are differences between objects floating in air and objects floating in water. There is no distinct "top" to the atmosphere, rather it becomes less dense with altitude. While a cork will float to the surface of water, a released helium-filled balloon rises until its weight equals the weight of the air displaced. Since air becomes less dense with altitude, the air that is displaced weighs less and less as the balloon rises.

Equation of Continuity

We next consider what happens when gases or liquids are in motion. For liquids, which are normally incompressible, the density of the liquid does not change under ordinary pressures. Then, the volume of liquid entering a pipe must equal the volume of liquid leaving it per unit of time. Otherwise, we would be violating conservation of energy, because mass would either be added to or subtracted from the system. The equation of continuity states that:

$$A_1 v_1 = A_2 v_2 \quad (see \quad Fig. \quad 7\text{-}6).$$

The product Av is constant throughout a pipe system of varying diameters. Where the cross-sectional area is large, the velocity is small, and vice versa. This is

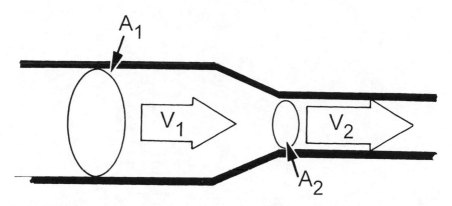

FIGURE 7-6. The liquid flows faster through a narrower pipe; the cross-sectional areas of the pipe are represented by A and the speeds of the liquid by V.

analogous to a river flowing slowly when it is broad, but speeding up when it passes through a narrow gorge. Similarly, one can apply this equation to the flow of blood in the human body. Blood flows from the heart into the aorta from which it passes into the major arteries and then into progressively smaller blood vessels, finally ending up in tiny capillaries. The cross-sectional area of the aorta is about 3 cm^2, and the blood flowing through has a speed of about 30 cm/s. There are a huge number of capillaries so that their aggregate cross section is about 2,000 cm^2. Applying the equation of continuity, we find that the speed of blood flow in the capillaries is about 30 cm/s$\times$3 cm^2/2,000 cm^2=0.045 cm/s. Thus, the blood flows far more slowly through the capillaries than through the aorta or the major arteries.

Bernoulli's Principle

When a fluid flows through a narrow constriction, its speed increases. Daniel Bernoulli (1700–1782), investigated how the fluid acquired the energy for this increase in speed. He reasoned that the extra kinetic energy was at the expense of a lowered internal pressure. If the fluid is flowing faster, the pressure from the surrounding fluid opposing its motion must be less. Thus, **Bernoulli's Principle** states:

Where the velocity of a fluid is high, the pressure is low, and where the velocity of a fluid is low, the pressure is high.

Bernoulli's Principle holds only for **streamline or laminar flow**. In laminar flow, the path taken by each bit of fluid does not change with time. Thus, the motion of a fluid in streamline flow can be represented by streamlines which are indicated by curved lines in Fig. 7-7. Streamlines are the smooth paths taken by fluid molecules

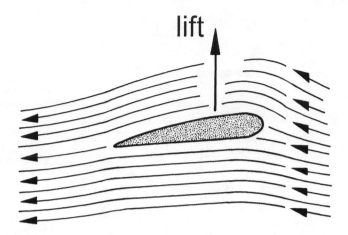

FIGURE 7-7. Streamline flow past a wing-shaped object. Since the fluid that moves above the wing travels a longer distance than that below it, the fluid speed above the wing is greater, resulting in a reduced pressure and a net upward force indicated by the arrow.

of neighboring regions which never cross one another. These lines are closer together in narrower regions where the speed of flow is greater and the pressure within the fluid is less. If the flow speed is too great, the flow may become turbulent and follow circular paths known as eddies. Bernoulli's Principle does not apply to such turbulent fluid flow.

There are many important applications of Bernoulli's Principle. Atmospheric pressure decreases in a strong wind such as a hurricane or tornado. Thus, a tornado passing over a house will create a partial vacuum above the roof of the house. If the pressure difference between the normal atmospheric pressure inside and the abnormally low pressure above the roof is too great, the greater pressure inside will blow the roof off the house.

Bernoulli's Principle also explains how airplanes stay aflight. As pictured in Fig. 7-7, streamlines above the curved airplane wing have to go a further distance than those below it. This means the air must move faster above the wing, and thus the pressure above the wing is less than the pressure below it. This results in a net pressure upward on the wing which lifts the airplane.

Airplanes also gain "lift" by virtue of air impact against the lower surface of the wings that are tilted back slightly for this purpose. These two contributions, the impact of the air plus Bernoulli's Principle, allow an airplane to leave the ground and keep it in the air. The amount of lift in both cases depends upon the relative speed of the plane with respect to the air—its airspeed. The faster the airspeed the greater the lift. For any given type of airplane there is a minimum speed needed to generate enough lift to leave the runway. Airports are designed with the proper runway distances that will be needed by the types of planes that will use them.

Another less dramatic example of Bernoulli's Principle is the functioning of chimneys. Air is in constant motion above the chimney, producing a somewhat lower pressure than the stagnant air below the chimney (inside the home). The greater air pressure inside the house then forces the air and the smoke from the fireplace up the chimney and out of the house.

In the film *Star Wars*, we see the star fighters attacking the Death Star at the end of the film. These little rocketships (whose rockets are located at and point out from the back of the ships only), make banked turns as they approach the enemy battle station (102 min). The motion of the star fighters looks perfectly natural, since it replicates the motion of fighter planes in combat on Earth. But real fighter planes can make these banked turns only because of air friction and Bernoulli's Principle. In the high vacuum of outer space which surrounds the Death Star, there would be no air friction and the rebel star fighters could not make banked turns while approaching their target. Instead, they would have crashed head-on into the Death Star, producing a rather unhappy ending to *Star Wars* and precluding any sequels.

In the beginning of *Blade Runner*, we see the hero of the film picked up by police who fly him in a "hovercraft" to their headquarters (10 min). It is not clear just how the hovercraft works. Air is thrust from the bottom of the hovercraft out of very small openings. The crafts are clearly not rocket-powered since they operate at low altitude within a city. Presumably, Bernoulli's Principle is applied to the rapid motion of blades on the underside of the hovercraft, producing a lift. Such small openings could only house small blades, however. How, then, do these blades produce enough lift for an entire car? The relative dimensions of the blades and the car seem to be wrong.

Diffusion

If a drop of coloring is placed in a pan of water, it spreads throughout the water until the color becomes uniform. This mixing occurs because of the random movement of molecules and is called **diffusion**. Diffusion also occurs in gases. Thus, perfume diffuses in air, although the motion of currents of air often plays a greater role in spreading odors than does diffusion. Generally, the diffusing substance moves from a region where its concentration is high to one where its concentration is low.

Diffusion is very important in life processes. Within cells, molecules produced in certain chemical reactions must diffuse to other areas to take part in yet other reactions. Gas diffusion is also important. For example, animals exchange oxygen and carbon dioxide with the environment. Oxygen is needed for energy-producing reactions within the cells, and thus must diffuse into them. Carbon dioxide is the result of many metabolic reactions and must diffuse out of cells.

Because of the slowness of diffusion over long distances, most animals have developed complex respiratory and circulatory systems. It can be shown that the time it takes for molecules to diffuse down a passage is proportional to the square of the distance over which they must diffuse. Thus, in the picture *Them!* the diffusion time for oxygen molecules to pass into the 100-in. giant ants through air passages in their sides would be 100^2 times longer than it would take the oxygen molecules to pass through the sides of a 1-in. ant. This 10,000-fold increase in the diffusion time would mean that the giant ants would suffocate, or at best, be extremely lethargic. In order to survive, they would need an entirely new breathing system, not just a scaled-up version of the branching system of tubes (tracheae) that leads from the body surfaces into the interior of small ants.

In the same film, cyanide gas grenades are hurtled into a giant ant nest in order to kill the insects (39 min). However, cyanide gas is 6% lighter than air, and thus would not quickly sink to the floor of the ant nest, and go down perhaps hundreds of feet through long tunnels into lower regions of the nest. Rather, it would diffuse slowly and penetrate the lower levels of the nest poorly, if at all.

Exercises

1. *Solid State*: Since a crystal has long range order, does this mean that two different pieces of the crystal would have a similar arrangement of atoms in them?

2. *Density*: How does mass density differ from weight density?

3. What is the mass and weight of 10 m³ of aluminum?

4. How many cm³ of ethyl alcohol will have the same mass as 15 cm³ of water?

5. *Size, Mass, and Strength*: Suppose that one could triple the size of a gorilla without changing the size of its body parts relative to one another. How many times stronger would the gorilla become, and how many times more would it weigh? Would it likely be able to perform feats of great strength like those shown in *King Kong*?

6. *Pressure*: If you are swimming at 3 ft below the surface of a lake and then you swim 9 ft below the surface, by what factor does the water pressure on your body increase?

7. If the surface of the water in a water tank is 20 m above the water faucet in a home, calculate the water pressure at that faucet.

8. In which floor of a home is the water pressure the greatest, the bottom floor or the top floor?

9. *Buoyancy*: If a rock is thrown into a deep pond, what happens to the buoyant force on it as it sinks deeper into the pond?

10. If a submerged fish makes itself less dense by puffing out its sides, what will happen to it?

11. Since the walls of a submarine are made of metal which is more dense than water, how can it ever float?

12. *Pascal's Principle*: If the pressure in a hydraulic press is increased by an extra 5 N/cm², how much extra weight will the output piston support if it has a cross-sectional area of 100 cm²?

13. *Atmospheric Pressure*: What is the approximate mass of a column of air 1 m² in cross-sectional area that extends from sea level to the upper atmosphere? What is the weight of this amount of air?

14. Convert a gauge pressure of 2 atmospheres to an absolute pressure of lb/in^2.

15. *Buoyancy of Air*: How much buoyant force acts on a dirigible that is floating in air and which weighs 10,000 N? Suppose that the dirigible loses some of its cargo. What would happen?

16. *Equation of Continuity*: How fast would blood flow through major arteries with a total cross-sectional area of about 2 cm^2? Use the data in this chapter for the speed and cross-sectional area of blood flow in the aorta.

17. *Bernoulli's Principle*: How might Bernoulli's Principle explain why gophers that live underground are able to avoid suffocation? Note that their burrows always have at least two entrances.

18. Why does an aircraft carrier normally turn into the wind before launching its airplanes?

19. Why does an umbrella sometimes turn inside out in a high wind?

20. Why might the cloth top of a convertible bulge out when the car is traveling at high speed?

21. *Diffusion*: In terms of the diffusion process for taking in oxygen and emitting carbon dioxide, explain why there have never been giant insects roaming the Earth.

Chapter Eight
Heat, Temperature, and Thermodynamics

Temperature

The description of heat and its transfer requires the establishment of a temperature scale. The statement that something is hot or cold is too vague to be of much use. Also, it can be confusing. For example, imagine that you have three pans of water lined up side by side. The left pan contains very warm water, the middle pan holds lukewarm water, and the right pan holds cold water. Place your left hand into the very warm water and your right hand into the cold water. Then place them both into the pan of lukewarm water. Your left hand will feel that the water is cold and your right hand will feel that it is hot. Which is correct? This illustrates the need for an objective way of measuring temperature.

The **temperature** of an object is a measure of how hot or cold the object is with respect to some set of standards. Many properties of matter change with temperature. Most materials expand when heated: that is why expansion joints are placed at regular intervals on the roadway of bridges, for example. The electrical resistance of materials also changes with temperature.

There are two temperature scales in common use in the US today. Both scales assign a given number to the temperature at which water freezes and another number to the temperature at which water boils. Table 8.1 below gives these numbers for the two temperature scales, the Fahrenheit scale and the Celsius scale.

Table 8.1. Temperatures in Degrees of Celsius and Fahrenheit

	Celsius (C)	Fahrenheit (F)
Water freezes	0	32
Water boils	100	212

Thus, the temperature interval between the freezing and boiling points of water can be stated as either 100 °C or 180 °F. Thus, 100 °C=180 °F or 1 °C=1.8 °F. The Celsius degree is thus larger than the Fahrenheit degree. One can readily convert temperatures from one system to another using the formula:

$$T \text{ °F}=1.8T \text{ °C}+32.$$

One can see that by substituting 100 for T °C the formula yields the correct answer for T °F, namely, $1.8 \times 100 + 32 = 212$ °F.

Suppose that the temperature of your home is 72 °F. What is 72 °F in degrees of Celsius? Substitute 72 for T °F in the conversion formula, which becomes 72 °F=1.8T °C+32, then 40=1.8T °C and thus 22.2=T °C.

There is a third temperature scale used by scientists, and that is the Kelvin (K), or absolute temperature scale. This is related to the Celsius scale in that their units are the same, only the point referred to as zero differs. Specifically,

$$T° \, K = T \, °C + 273.$$

Thus the zero on the Kelvin scale is -273 on the Celsius scale. As we will see later, it is the point at which all of the kinetic energy which can be removed from a substance has been removed. This temperature is called **absolute zero**.

Temperature is a measure of the *average* kinetic energy of the atoms or molecules in a substance, not a measure of the *total* kinetic energy of the molecules and atoms. Thus, 10 kg of boiling water has 10 times the molecular kinetic energy of 1 kg of boiling water. Yet the temperatures of the two quantities of water are identical, because the average kinetic energy per molecule of water is the same in both quantities of boiling water.

A **thermometer** is an instrument designed to measure temperature. The most common type is a hollow glass tube filled with colored mercury or alcohol. A bulb at the bottom of the tube acts as a reservoir for the liquid. The thermometer records the relative expansion or contraction of the glass tube relative to the liquid inside of it. Since the liquid expands more than the glass with an increase in temperature, it rises up the tube as the temperature of the tube increases. It should be noted that thermometers measure their *own* temperature, so one must place the thermometer in good thermal contact with the object whose temperature you wish to measure so that the temperature of the object and the thermometer are the same.

Another common temperature dependent instrument is the bimetallic strip, depicted in Fig. 8-1. It consists of two dissimilar metals welded together. Suppose that the strip is flat at room temperature. When the temperature rises, the two metals expand differently and the strip bends as indicated. A contact at the end of the bottom strip could activate an electrical circuit which would, e.g., turn on an air conditioner. When the temperature cools down the strip again becomes flat and the air conditioner is turned off. Bimetallic strips are used in temperature controls on such common household appliances as refrigerators, ovens, electric toasters, and home heating units.

Thermal Expansion

Most substances expand when heated and contract when cooled. It is important that materials in close contact with each other expand at the same rate when exposed to temperature increases. Thus, the material used in a dental filling must expand at the same rate as the tooth it is placed into, or the filling will fracture or even fall out as the tooth is subjected to temperature change.

One exception to this general rule (that most substances contract when cooled) is water. At about 4 °C water occupies its smallest volume. It is thus densest at that

bimetallic strip

room temperature

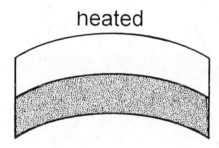

heated

FIGURE 8-1. A bimetallic strip: the metal on top expands more than that on the bottom with an increase in temperature.

temperature. At both lower and higher temperatures a given amount of water occupies more space, i.e., is less dense than the water at 4 °C. This peculiar behavior is due to the fact that ice crystals start to form at about 10 °C, and the formation of more of them as the temperature decreases from 10 to 4 °C offsets the "normal" contraction that water would experience as its temperature is lowered below 10 °C. Below 4 °C the formation of ice crystals predominates as the temperature is further lowered so that the volume of water increases with decreasing temperature. If we lower the temperature to 0 °C, all of the water will eventually freeze and ice occupies about 10% more volume than the same mass of water. This property of water is important in nature. It means that when a lake or pond is cooled in the winter, before freezing occurs, all of the lake must first be cooled to 4 °C. Then the water nearest the surface can be cooled further until it reaches 0 °C and freezes. Meanwhile, the densest water in the lake is at 4 °C and lies at the bottom of the lake. Fish and plants survive the winter in this 4 °C water, since the ice that forms at the top of the lake is a poor thermal conductor and thus prevents the lower levels from giving up further heat and becoming cold enough to freeze.

Heat

One can define **heat** as a transfer of energy from one body to another that occurs because of a difference in temperature between the bodies. In the SI system of units, the unit for heat is the same as for any form of energy, namely, the joule. Another common unit for heat is the **calorie**, which is defined to be the amount of heat necessary to raise the temperature of 1 g of water by 1 °C, from 14.5 to 15.5 °C. This temperature range is specified because the amount of heat required varies (by less than 1%) over the temperature range 0–100 °C. We shall ignore this variation in our discussions. One calorie can be shown to be equivalent to 4.18 J. The reader should be aware that the calorie of physics is different from the calorie of food. The food unit is sometimes written Calorie and it equals 1,000 calories as defined above. Dieters would soon starve on a daily intake of 1,500 physics calories!

Specific Heat

You have probably noticed that some foods remain hotter longer than others. We describe this property of matter by saying that those foods have a higher specific heat than others. **Specific heat** can be defined as the quantity of heat required to change the temperature of a unit mass of the substance by 1 °C. The higher the specific heat the more heat is needed to raise the temperature of a unit mass of the material. Water has by far the highest specific heat of any common material. For example, iron's specific heat is only 11% that of water. The human body, which contains a great deal of water, has a specific heat 83% that of water.

The high specific heat of water is why it is used in the cooling system of automobiles. It can absorb a great deal of heat but only experience a small rise in its temperature. Water also takes a long time to cool, which explains why hot water bottles were once frequently used on cold nights. It also explains why climates are milder near the oceans. Because of the huge mass and high heat capacity of the oceans, an ocean does not vary much in temperature from winter to summer. Islands and peninsulas that are surrounded by water do not have the extremes of temperatures that occur in the interior of continents.

Heat Transfer

Heat is transferred from hot to cold objects by three mechanisms: conduction, convection, and radiation. **Conduction** is the movement of heat from molecule to molecule of a substance. How well a substance conducts heat depends upon the electrical bonds between its molecules. Generally, good electrical conductors such as metals are good heat conductors. Good electrical insulators are generally poor conductors of heat. Thus pots and pans are made of metal, while wood is used for the handles of pans because it is a poor heat conductor.

Liquids and gases are normally poor heat conductors. Thus, snow, which contains many trapped air spaces, is a good insulator. Similarly wool, fur, and feathers contain large trapped air spaces and are also good insulators. You feel warmer in a feather-filled coat on a cold winter day because the coat keeps your body heat from escaping. The coat does not keep the cold out; it keeps the heat in.

Convection is heat transfer by the actual motion of fluids. It can occur in any liquid or gas. Let us apply this concept to the rising of warm air. We could explain it using the concept of buoyancy: warm air expands and is less dense than the colder air which then exerts a buoyant force that pushes the warm air upward.

Radiation is the name given to the transmission of heat via electromagnetic waves. This is the way energy reaches the Earth from the Sun. Some of this energy is visible, other components such as the ultraviolet and short-wavelength infrared radiation are invisible to our eyes, although we can feel their warmth when absorbed by our body. At ground level in the US we receive about 180 W/m^2 of radiant energy from the Sun. Most of the Sun's radiant energy passes through the atmosphere and is absorbed at the surface of the Earth. The Earth then reradiates energy back toward space mainly in the range of longer-wavelength infrared. Much of the Earth's radiant energy is absorbed by carbon dioxide and water vapor in the atmosphere which then reradiates it back to the Earth. This process is called the **greenhouse effect** and results in the Earth being at a higher temperature then it would be without that effect. There is concern that pollutants emitted into the air may increase the greenhouse effect, leading to global warming that could dramatically change the climates of major parts of the world, and not necessarily for the better.

Science fiction films have dealt with the devastating results of pollution and the greenhouse effect. *Silent Running* describes a bleak future in which all vegetation on Earth has died and the last plants are on domed greenhouses attached to space freighters (6 min) in orbit around Saturn. This plot premise does not make sense. It would take an enormous amount of energy to move these greenhouses to Saturn. Also, the plants would be receiving much less radiant energy from the Sun than if the greenhouses were placed in orbit around the Earth. The average radius of the orbit of the Earth about the Sun is 93 million miles while the average radius of the orbit of Saturn is 887 million miles. The intensity of the sunlight falls off as one over the square of the distance from the source, the Sun. Thus the intensity of sunlight at Saturn would be only $(93/887)^2 = 0.01$ of the intensity of the sunlight at the distance of the Earth from the Sun. This would be insufficient to support the life of most of the plants in the greenhouse.

There is, in addition, the danger of the greenhouses in space being struck by meteorites. It is hard to see how they could be adequately protected against that danger.

Science fiction has also dealt with very cold environments. In *The Empire Strikes Back*, we see a vivid example of a way of surviving at extremely cold temperatures. At the beginning of the film (4 min), Luke Skywalker is attacked by a huge bearlike creature on a snow- and ice-covered planet. Han Solo reaches him as Skywalker collapses in the snow. Solo's mount, an alien beast called a tonton, dies from the cold and Solo is left with the predicament of how to survive a horrendously cold night. The solution is very reasonable: he cuts open the dead tonton, whose insides appear to contain a good deal of water and intestines, and he stuffs the unconscious Skywalker inside the beast (13 min). The high specific heat of the tonton keeps Skywalker from freezing to death while Solo constructs an igloo from the snow and ice. Because of the insulating properties of snow described above, the structure keeps the two men alive until they are rescued the next morning.

In *Phase 4*, a research station is attacked by ants which concentrate solar reflectors against the station, thus raising its temperature substantially and causing the computers to malfunction (44 min). But the reflectors are too small to achieve that result. At best, the reflectors could direct 100% of the radiant energy they received from the Sun against the research station. Since the combined cross-sectional areas of the reflectors were at most 10 m^2, they could have directed, at most 10×180 W/m$^2 = 1,800$ W against the center. Some of this would have been reflected by the walls of the research center; the amount absorbed would not have been sufficient to raise the temperature of such a large structure 5° in 1 1/2 hours. One of the scientists in the film recognizes this contradiction by saying that the structures must be directing more than sunlight toward the facility, but there is no evidence for any other source of the bright light coming from the reflectors except sunlight.

In *2010*, the alien monolith aids the development of life on the cold surface of Europa, a moon of Jupiter, by turning Jupiter into a second sun (105 min). Even though the energy output of that second sun was less than that of our solar system's Sun, it was so much closer to Europa that the end of the film depicted water, rather than ice, on the surface of Europa (111 min).

Change of State

Matter can change from the solid to the liquid, and then to the gaseous state by the addition of heat. Melting is the changing of matter from the solid to the liquid state. As energy is added to a solid, the atoms vibrate more and more until the bonds holding them in a rigid lattice are broken and they are able to move freely over one another. The atoms are then in the liquid state. The melting of a substance occurs at constant temperature since the added heat all goes into breaking the rigid lattice. The amount of heat needed to melt 1 kg of a substance is called its **heat of fusion**. For water the heat of fusion is 80 kcal/kg.

The heat required to change a kilogram of liquid to a gas is called the **heat of vaporization**. For water, this is 540 kcal/kg. The heat gives individual water molecules enough kinetic energy to break loose from the water and to become a gas.

Figure 8-2 shows the heat needed to raise the temperature of 1 kg of ice from -20 °C to 0°, melt it, raise the water to 100 °C, and then boil off the water. The

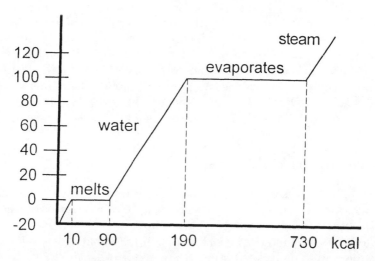

FIGURE 8-2. The amount of heat needed to raise the temperature of 1 kg of ice and to change its state to liquid and to gas.

heat added in each of the changes of phase can be extracted by reversing the process. Thus, when steam in radiators condenses back into water, it releases 540 kcal/kg of heat energy. This is why steam heat was used in many homes; the high heat of vaporization made the steam an effective way of carrying heat from a furnace throughout a home.

Generally, evaporation of a liquid cools the object in contact with it, and condensation of a liquid warms the object it condenses upon. Thus, sweating cools the body as the sweat evaporates. In *Them!*, the scientists correctly state that ants forage in the desert only at night (35 min). This is because the high temperatures during the day are very uncomfortable for them, since they cannot perspire to lower their body temperatures. Some animals, such as dogs, cool themselves by panting. Evaporation occurs in their mouths and bronchial tracts. In dry climates, evaporation is greater than condensation so that you feel cooler at a given temperature in a dry location such as Arizona rather than in a humid location such as Florida.

The evaporation of a liquid can be used to cool food and water. Porous canvas bags filled with water can be cooled by hanging them outside of a car when it is moving quickly. Some water molecules will seep through the canvas and evaporate, cooling the rest of the water. Similarly, wrapping a bottle with a wet cloth can produce as much cooling as placing it into a bucket of cold water because of the cooling effects of the evaporation of the water in the cloth.

In *The Thing*, the block of ice in which the alien is encased when it is discovered by the expedition should not have melted so quickly (33 min). There was an open window into the storeroom in which the body was placed: the room was bitterly cold with its temperature below freezing. The heating element on the blanket might have melted some ice directly under it, but the water would quickly freeze again. The film instead depicts the entire top-half of the ice block, which is 8-ft-long by at least 2-ft-wide and 3- or 4-ft high, as having melted during part of the 2 hour shift of one of the guards. The heat needed to melt half of this block is huge. The volume of ice to be melted is perhaps 1 m^2 of ice, which would have a mass of about 1,000 kg and melting that amount of ice would require about 80,000 kcal or over 3.2×10^8 J of energy. If the blanket provided 1,000 W of energy it would take nearly 100 h to produce the needed amount of heat!

Thermodynamics

Thermodynamics can be defined as the study of processes in which energy is transferred as heat and/or as work. We will apply this to **systems**, namely, any set of objects we wish to consider. A **closed system** is one in which no mass enters or leaves. An **open system** is one in which mass may enter or leave. A closed system is said to be **isolated** if no energy enters or leaves it.

The First Law of Thermodynamics

The **First Law of Thermodynamics** is simply the application of the law of conservation of energy to thermal processes:
Heat added to a system=external work done by the system+change in its internal energy.

Any heat added to a system is either transformed by the system into external work or it remains within the system. We call the energy within any system of objects its **internal energy**. This energy has many components, including kinetic energy, potential energy, and even the energy inherent in mass itself (see the chapter on relativity for Einstein's famous $E = mc^2$). Suppose that 1,000 J of heat is added to a system that does no work. Then its internal energy must increase by 1,000 J or the process would violate the conservation of energy law.

There are some processes in which no net heat enters or leaves a system. We call such a process an **adiabatic process**. One can approximate adiabatic processes by isolating a system from its surroundings using thermally insulated materials, or by performing the process so quickly that heat has very little time to enter or leave. If a gas expands adiabatically, since no heat enters or leaves, the First Law of Thermodynamics implies that the work done by the gas in expanding must be derived from the internal energy of the gas. If this internal energy decreases, the gas molecules move more slowly on average and thus have a lower average kinetic energy. Since the temperature of the gas is a measure of the average kinetic energy of the gas molecules, the temperature of the expanding gas will decrease. This general conclusion is an example of why the laws of thermodynamics are so powerful: They allow one to make predictions for the behavior of systems that are independent of the details of the processes involved.

One example of an adiabatic process is the compression and expansion of the pistons in an automobile engine. These occur in only hundredths of a second, too short a time for appreciable amounts of energy to leave the combustion engine.

Second Law of Thermodynamics

There are many processes which would not violate the First Law of Thermodynamics, yet they never occur. For example, if you place a metal teaspoon in a cup of hot coffee, the temperature of the coffee decreases and that of the spoon increases until they are both at the same temperature. Heat flows from the coffee to the spoon. It never flows from the spoon into the coffee, making the coffee even hotter and the spoon icy cold. This one-way heat flow, from hot to cold, is called the *Second Law of Thermodynamics*:

Heat flows naturally from a hot object to a cold object; heat will never of itself flow from a cold object to a hot object.

The useful work produced by a heat engine (see Fig. 8-3) is the result of heat flowing from a high temperature energy reservoir to a low temperature energy reservoir while part of the heat is converted into work. In a gasoline engine, for example, the fuel which burns in the combustion chamber is the high temperature energy reservoir, the pistons perform work, and heat is emitted out the exhaust pipe of the car to the surrounding air, which is the low temperature energy reservoir. The second law can be restated in terms of heat engines:

No heat engine is possible whose sole effect is to transform a given amount of heat completely into work.

Thus, there is always some heat exhausted from such an engine. The fraction of heat that can be converted into work is theoretically limited by the relative temperatures of the two heat reservoirs between which the engine operates. Sadi Carnot (1796–1832) proved that the most ideal engine conceivable (one that took an infinite time to complete one cycle in order to minimize internal friction of the working substance) would have an ideal efficiency (defined to be the ratio of the work done by the engine divided by the heat put into the engine) given by:

$$\text{efficiency} = (T_h - T_c)/T_h ,$$

where T_h is the Kelvin temperature of the hot reservoir and T_c is the Kelvin temperature of the cold reservoir. Thus, an engine operating between the boiling and freezing points of water, 100 and 0 °C has an efficiency $= (373-273 \text{ K})/373 \text{ K} = 0.27$ or 27% of the heat entering this engine could be changed into work. One

heat engine schematic

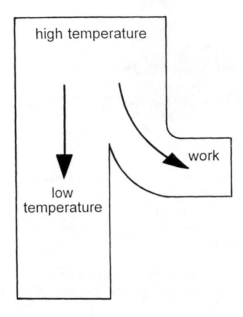

FIGURE 8-3. A schematic of a heat engine.

could have a 100% efficient engine only if $T_c = 0°$ K. Since no further energy can be removed from the molecules, the temperature cannot be lowered further. No engine can operate between some temperature and absolute zero, since the contact with absolute zero would result in some energy being added to it, and the temperature would thus not be absolute zero. It is an unattainable temperature, and thus a 100% efficient engine is unattainable.

There is an alternative way of interpreting the Second Law of Thermodynamics. Since energy flows only from high temperatures to lower temperatures, this flow degrades the energy in the sense that it is less useful at lower temperatures because it can be used to run fewer heat engines and those will have to operate using less of a temperature difference between the two heat reservoirs. Further, the energy goes from a more organized form, such as gasoline, to a less organized form, such as the emissions from an automobile. Energy in the form of electricity that flows into homes and offices is converted into heat energy in the light bulbs. One could say that heat is the graveyard of useful energy. One may state the Second Law of Thermodynamics in yet another way:

Natural systems tend to move to a state of greater disorder. (By disordered we mean more random.)

If one has a bottle of salt and a separate bottle of pepper, it is a more ordered state than if one mixes the two together in a larger bottle because one knows that any given salt molecule must be in the smaller salt bottle. Once mixed with pepper, the location of the salt molecule is less certain, because it could be anywhere in the larger volume of the larger bottle. The term **entropy** is used to measure the amount of disorder in a system. Not every system grows more disordered; living things tend to evolve into more advanced (more ordered) creatures. But this is done at the expense of the rest of the universe; the waste products of living things increases the entropy of the rest of the universe more than the decrease in entropy that occurs within living creatures. This fact, that the entropy of the universe is constantly increasing as the universe moves to a state of greater disorder whose energies will be found at progressively lower temperatures, leads to the prediction called the **heat death of the universe**.

In the film *Zardoz*, a futuristic science has tried to overcome the increase in entropy with time by isolating a small section of the planet (called a vortex) from the rest of Earth (66 min). Within the vortex, death is banished through the use of advanced technology for a small group of humans who try to evolve into demigods. Thus, inside the vortex, entropy could be viewed as decreasing rather than increasing. However, the rest of the planet is descending into another Dark Age: one could consider that the entropy (disorder) of the Earth as a whole continues to increase even though it decreases within the vortex.

Exercises

1. *Temperature*: The average normal temperature of the human body is 98.6 °F. What is this temperature in degrees Celsius?

2. The temperature of a filament in a light bulb is about 1800 °C. What is this on the Fahrenheit scale?

3. *Thermal Expansion*: Suppose that you heat a metal ring. Does the size of the hole increase, decrease, or stay the same?

4. Steel expands about 1 part in 100,000 per °C increase in temperature. Suppose that a 1 km bridge had no expansion joints in it. How much would it expand if the temperature increased by 10 °C?

5. *Heat*: Suppose that you consume 1,000 food Calories and that you have a mass of 50 kg. Further suppose that you consist essentially of water. How high an increase in your body temperature could be provided by the 1,000 Calories if all of the food energy went solely into raising your temperature?

6. *Specific Heat*: Bermuda is close to North Carolina. Why does Bermuda have a tropical climate year-round, but North Carolina does not?

7. *Heat Transfer*: How does a blanket keep you warm on a cold night, since it is not a source of energy?

8. Why do restaurants often serve baked potatoes wrapped in aluminum foil?

9. *Change of State*: Why does blowing over hot soup cool the soup?

10. Why do farmers spray fruit trees with water to help prevent the fruit from freezing if an unusually late frost is forecast?

11. *First Law of Thermodynamics*: In an adiabatic process, if work is done **on** a system (such as by compressing a gas enclosed in a cylinder), what happens to the internal energy of the system?

12. If you shake a can of liquid back and forth quickly, will its temperature rise?

13. *Second Law of Thermodynamics*: Why will a refrigerator with a given amount of food consume more energy in a warm room than in a cold room?

14. What is the efficiency of a heat engine operating between 123 and −73 °C?

15. What happens to the efficiency of a heat engine if one lowers the temperature of the reservoir into which the heat is emitted?

Chapter Nine
Wave Motion and Sound

The concept of waves is of central importance to an understanding of the world around us. Energy may be transmitted via a wave or through the kinetic energy of a moving particle. Sound and light are both vibrations that move through space as waves. Sound propagates only through the vibrations of a material medium, be it solid, liquid, or gas. If there is no medium to vibrate, it is impossible to transmit sound. Light, on the other hand, is a vibration of electric and magnetic fields and can pass through material media or empty space. However, the source of all waves, light or sound, is something that vibrates.

Waves

Figure 9-1 depicts a typical wave motion. The wave shown is a sine curve. The **amplitude** is the maximum displacement of the wave from its position of rest. The **wavelength** is the distance from the top of one crest to the top of the next. The **frequency** of the wave specifies the number of complete cycles per second undergone by the source of the wave. This is equivalent to the number of wavelengths that pass a given point per second. For example, the frequency of a vibrating pendulum specifies the number of complete to-and-fro motions it makes in 1 s. If three such complete vibrations occur in 1 s, the frequency is said to be 3 vibrations/s, or 3 Hz. Higher frequencies are measured in kilohertz, where 1 kHz is 1,000 vibrations/s. Still higher frequencies are in megahertz (MHz), where 1 MHz is 1,000,000 vibrations/s. AM radio waves are in the kilohertz range while FM radio waves are in the megahertz range. These radio wave frequencies are the oscillation rates at which electrons vibrate in the antenna of a radio station's transmitting tower.

The **period** is the length of time it takes for a complete vibration. The period is equal to 1/frequency. For example, if the frequency is 3 Hz then the period is one-third of a second.

Wave Speed

The speed of a periodic wave motion is equal to the frequency times the wavelength of the wave. It is easy to see that this must be the case by imagining water waves moving past a given point. If 10 water waves move past the point per second, and the distance between each succeeding water crest is 1 m, then 10 m of the wave passes the given point in a second. The frequency in the above example is 10 Hz, and the wavelength is 1 m. Thus, wave speed equals 10 Hz × 1 m = 10 m/s.

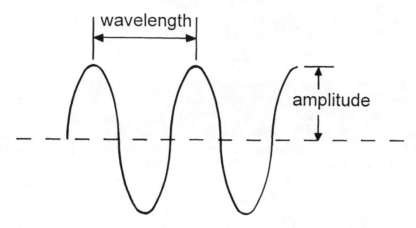

FIGURE 9-1. The wavelength and amplitude of a wave.

As a further example, consider a water wave that oscillates up and down five times each second with a distance between wave crests of 2 m. In that case the frequency is 5 Hz; its wavelength is 2 m, and its wave speed is $5 \times 2 = 10$ m/s.

Transverse Waves

There are two types of waves. If the motion of the medium is transverse to the direction of the wave motion, the wave is called a **transverse wave**. The top half of Fig. 9-2 is an example of a transverse wave moving through a long spring. Another example is to fasten one end of a rope to a wall and shake the free end up and down. A pulse will travel along the rope and return. In this case, the motion of the rope is up and down, i.e., vertical, while the motion of the wave is horizontal.

Longitudinal Waves

If the particles move along the direction of the wave rather than at right angles to it, we call this a **longitudinal wave**, as pictured in the bottom half of Fig. 9-2. Another example of longitudinal waves is sound in air. The air molecules vibrate to and fro along the direction the sound wave moves.

Interference

Unlike material objects, more than one wave can occupy the same space at the same time. If two or more waves occupy the same space at the same time the result is referred to as **interference**. As the top-half of Fig. 9-3 indicates, for the

transverse wave

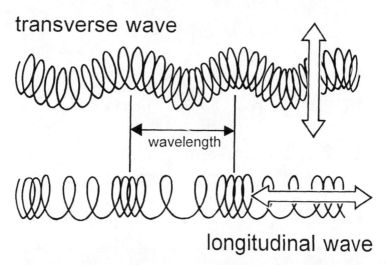

wavelength

longitudinal wave

FIGURE 9-2. A longitudinal wave (bottom spring) and a transverse wave (top spring).

superposition of two identical transverse waves, if the peaks of both waves coincide with one another, their individual effects add together, and we have a wave of increased amplitude. This is called **constructive interference**.

When the wave crest of one wave coincides with the trough of another, the waves cancel. This is called **destructive interference**, as pictured in the bottom-half of Fig. 9-3, in which the solid straight line is the resultant of the two superimposed waves represented by the dashed lines. Interference is a characteristic of all wave motion, whether the waves are water waves, sound waves, or light waves.

The Doppler Effect

If either the source of the waves or the observer is moving relative to one another, the distance between successive wave crests reaching the observer is affected. For example, if a fire engine has its siren on as it approaches you, your ear hears a higher frequency until it gets alongside of you, and then its frequency drops suddenly as it moves away from you. This is because the siren is moving as it emits sound waves. It therefore pushes the sound waves closer together in the direction it moves. You hear this decreased distance between sound waves as a higher frequency, since frequency is proportional to the number of wave crests passing your ear per second. The moment the fire engine passes you, the distance increases between successive sound waves reaching you, and you, therefore, hear this as a lower frequency, as indicated in Fig. 9-4. Similarly, light emitted by a source moving away from the observor would be shifted to longer wavelengths (the so-called "redshift").

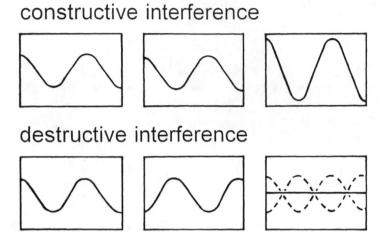

FIGURE 9-3. Two transverse waves demonstrate constructive interference in the top-half and destructive interference in the bottom-half of the figure.

Sound

As you may have concluded, many of the sounds that you hear are produced by something vibrating. The "something" may be the vocal chords of a friend, a vibrating string, or a vibrating column of air in a flute. In each of these cases, the vibrating source sends a longitudinal wave through the surrounding air to your ears. Sound waves with frequencies below 20 Hz are called **infrasonic waves** and those with frequencies above 20,000 Hz are called **ultrasonic waves**. The human ear can only hear sounds with frequencies between about 20 and 20,000 Hz. Dogs and other animals can hear sounds over a wider range of frequencies than can humans.

In *Superman*, the villian, Lex Luthor, calls Superman by using a sound frequency greater than 20,000 Hz (95 min). Luthor tells the Man of Steel that he is the only creature not walking on four legs that can hear this message. Luthor is right!

Speed of Sound

The speed of sound depends on the medium it is traveling through. The speed of sound in dry air at 0 °C is about 331 m/s. The speed increases with increasing temperatures by about 0.6 m/s/°C. Thus, at a normal room temperature of 20 °C, the speed of sound is about 343 m/s. In water sound travels several times faster than in air, and in metals its speed is even greater.

Some animals emit ultrasound waves (i.e., waves at frequencies higher than 20

FIGURE 9-4. The frequency of sound emitted from a fire truck is heard to decrease as it passes an observer because the distance between successive sound wave crests increases.

kHz) and locate objects by the time it takes the sound to move from the animal to the object and back again. For example, the ultrasonic waves emitted by a dolphin enable it to locate fish.

Sonar ("sound navigation ranging") is an invention which uses pulses of sound waves to locate objects in water. Here the sound waves have frequencies in the range of human hearing. It was invented during World War II to locate enemy submarines. The sonar unit measures the length of time between the emission of a pulse of sound waves and the detection of its reflection. The distance traveled by the sound waves in water during the intervening period is equal to the speed of sound in sea water times the time elapsed. Dividing this distance by 2 determines the position of the object reflecting the sound waves.

In *The Abyss*, the American submarine which is following the alien creature/ship at the beginning of the film uses sonar to track its target (1 min). In addition, sonar is used by the submarine to determine the distance to the walls of the undersea canyon that it traverses. Later in the film sonar is used to locate the sunken submarine and sense the movement of the alien undersea city at the end of the film (130 min). Sonar could not be used to locate the lost missile because it was too small to reflect waves back with sufficient intensity to be detected.

In *The Day of the Triffids*, the plants locate their victims by sound. If a potential victim makes no sound, the plants do not know the victim exists. When the hero of the film determines that sound attracts the triffids toward potential prey, he is able to lead them away using a truck equipped with a loud speaker system (83 min). While sound is used by animals, there is no indication that plants have this type of sensory capability.

Resonance

Any object composed of an elastic material will vibrate at its own characteristic frequencies. We know, of course, that tuning forks vibrate at their characteristic frequencies. So also do atoms and many other objects. We speak of these frequencies as being the object's natural frequencies. The object will absorb energy from an external wave oscillating at these natural frequencies and will vibrate with increasingly greater amplitude. This dramatic increase in the amplitude of the object's motion is called **resonance**. One example of this is a singer whose voice can shatter glass. Another example is a swing pumped in rhythm with its natural frequency. This produces a very large amplitude in the motion of the swing. A third example occurred in 1940, when the Tacoma Narrows Bridge in Washington state was destroyed by a resonance generated by the wind. The motion of the bridge steadily increased in amplitude until this mile-long bridge collapsed. For the same reason, troops marching across bridges are told to break step in order to prevent the possible collapse of the bridge if their cadence were identical to a natural frequency of the bridge.

Loudness

The intensity of sound depends on pressure variations within the sound wave. **Loudness** is a psychological sensation of sound. If you turned up a radio until it seemed about twice as loud as before, you would have to increase the power output, and therefore the intensity, by approximately a factor of 8. Human ears are not very good at judging loudness. The unit for measuring the relative loudness is the **decibel**. The threshold of hearing has an intensity level of 0 dB. The rustle of leaves is about 10 dB, a quiet radio about 40 dB, a siren at 30 m about 100 dB, and a loud indoor rock concert is about 120 dB, the threshold of pain. However, one's hearing can be damaged by exposure to much lower intensity sounds. One should note that the decibel scale is logarithmic, i.e., an increase of 10 dB means an increase in the intensity of sound by a factor of 10.

Shock Waves and the Sonic Boom

An airplane traveling faster than the speed of sound is said to have **supersonic speed**. Such a speed is often described by a **Mach number**, which is defined as the ratio of the speed of the object to that of sound in the medium at that location. For example, a plane traveling 900 m/s, where the speed of sound is 300 m/s, has a speed of Mach 3.

TABLE 9.1. Electromagnetic Spectrum.

Name of Wave	Wavelengths (m)
Radio waves	$3 \times 10^4 - 3$
Microwaves	$3 - 3 \times 10^{-4}$
Infrared	$3 \times 10^{-4} - 4 \times 10^{-7}$
Visible light	$4 \times 10^{-7} - 7.5 \times 10^{-7}$
Ultraviolet	$7.5 \times 10^{-7} - 10^{-8}$
X rays	$10^{-8} - 10^{-11}$
Gamma rays	$10^{-11} - 10^{-14}$

As an object moves faster than the speed of sound, a **shock wave** occurs. This results from the fact that when the object is traveling at the speed of sound, the wave crests it emits in the forward direction pile up directly in front of it. These different wave crests overlap one another to form a single large crest, called a shock wave. A shock wave in air is analogous to the wave of a boat traveling faster than the speed of the water wave it produces.

When an airplane moves at supersonic speeds the shock wave contains a great deal of energy and is heard as a "sonic boom." A sonic boom has enough energy to break windows and cause other damage. It should be noted that a shock wave occurs whenever an airplane is traveling at supersonic speed, not only when it first exceeds the speed of sound.

In *The Day the Earth Stood Still*, the alien spacecraft is reported to be moving at 4,000 miles/h and at a height of 20,000 ft toward Washington D. C. (2 min). Since the speed of sound in dry air at 0 °C is about 740 miles/h, this corresponds to a supersonic speed that certainly would have resulted in a sonic boom. Yet, no such boom is heard as the spacecraft lands in Washington

The Electromagnetic Spectrum

Light and other electromagnetic waves will move through both material media and a vacuum. These waves are called **transverse waves** since the electrical field vectors and the magnetic field vectors both oscillate at right angles with respect to each other and with respect to the direction of the wave motion. In a vacuum all electromagnetic waves move at the same speed, 3×10^8 m/s, and differ from one another only in their frequencies and wavelengths. As indicated in Table 9.1, one can catalog electromagnetic waves by their wavelengths. In Table 9.1 the range of wavelengths is approximated for each type of wave. Visible light has wavelengths in the range $4 \times 10^{-7} - 7.5 \times 10^{-7}$ m. One can determine the corresponding range of frequencies by dividing the speed of light, 3×10^8 m/s by the wavelength. This yields for the frequencies of visible light $0.75 \times 10^{15} - 0.4 \times 10^{15}$ Hz.

The color of an object is determined by the predominant wavelength of light emitted by, or reflected from the object. Violet light has a wavelength of about

4×10^{-7} m while red light has a wavelength of about 7×10^{-7} m. Passing white light through a prism separates it into its constituent colors, from violet to red. The rainbow is caused by the separation of white light by water droplets (which act like minute prisms) into its full spectrum. This phenomena is called **dispersion**.

X rays and gamma rays have much shorter wavelengths than visible light, while infrared, microwaves, and radio waves have much longer wavelengths. The great differences in the size of the wavelengths result in very different properties for these types of electromagnetic phenomena. For example, gamma rays can penetrate inches of lead shielding, while radio waves cannot penetrate even thin metal sheets.

In *The Day of the Triffids*, the triffids were apparently mutated by some form of world-wide radiation emitted during a meteor shower (3 min). Simultaneously, most of the human population were rendered blind, presumably by the same meteor-shower radiation. While it is true that ultraviolet radiation can cause blindness by producing an inflammation of the external eye parts followed by scarring of the cornea, the eye doctor in the film states that the optic nerve is "gone" due to the meteor-shower glare (19 min). Further, the onset of blindness is not immediate but delayed. It is hard to imagine any type of electromagnetic radiation that would both blind the population and cause the simultaneous mutation of the triffids world-wide.

Light

The propagation of visible light is governed by fairly simple physical laws. The **Law of Reflection** states that when light is reflected from a surface between two mediums, the angle of incidence is equal to the angle of reflection as pictured in Fig. 9-5. A light beam may also pass from one medium into another, in addition to being reflected from the surface between them. If the speed of light differs in the two media, the light ray that passes into the second medium (called the **refracted ray**) will bend at an angle different than that of the incident ray, as pictured in Fig. 9-6. The angle of incidence, the angle of reflection, and the angle of refraction are all measured with respect to the normal to the surface separating the two mediums (see Figs. 9-5 and 9-6, the dashed lines represent these normals).

It is because of refraction that a number of common optical illusions exist. For example, a person standing in waist-deep water appears to have shortened legs from the viewpoint of an observer standing alongside of the pool or beach. The observer's eye assumes the rays that travel from the surface of the water to the viewer originates along a straight line path, as pictured in Fig. 9-7, whereas actually the ray is bent at the surface of the water. The refraction of light that passes from one medium to another is called **Snell's Law**.

The Law of Reflection explains how a flat (plane) mirror creates an image by the reflection of light from the mirror. This image is referred to as a **virtual image** since the reflected light rays do not actually originate from the image. Figure 9-8 indicates how a pencil would appear when viewed through such a plane mirror.

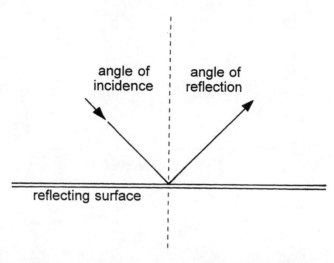

FIGURE 9-5. The angle of incidence is equal to the angle of reflection for a light ray striking a reflecting surface.

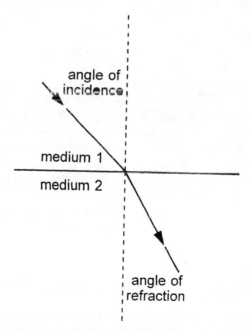

FIGURE 9-6. The angle of refraction differs from the angle of incidence if the speed of light in medium two is different from the speed of light in medium one.

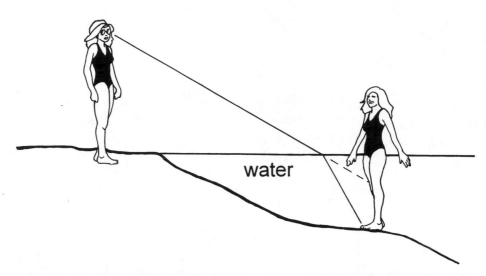

FIGURE 9-7. A person on the beach viewing someone waist deep in water will see the person as shorter than she really is.

Humans interpret light reaching their eyes as moving along a straight-line path from a light source. Thus, the dashed lines indicate the paths that the viewer thinks the light has taken from a virtual image of the pencil which appears to be behind the mirror. Solid lines indicate the actual paths taken by the light going from the pencil to the mirror and outward to the viewer's eyes. The Law of Reflection and Snell's Law have been used to construct lenses for a variety of purposes, from eyeglasses to the billion-dollar Hubble space telescope.

The lenses in our eyes also function according to these laws. The lens of an eye forms an image which will be in focus if it is centered on the retina (which is located at the back of each eye). The image is then transmitted via the optic nerve to the brain. When an observer views a distant object, nearby objects will appear to be out of focus. These nearby objects form images that would lie behind the retina and therefore their images on the retina are not sharp. This phenomenon occurs because the location of an image for an object at a given distance depends on both the distance between the object and the lens of the eye and on the focal length of the lens. Muscles in the eye change the focal length of the lens of the eye according to the distance of the object being observed. If that focal length is not the proper one for a closer object, then this closer object will appear fuzzy.

Polarization

Light waves are transverse waves since the electric and magnetic fields both oscillate at right angles to the direction in which the light ray moves. If the motion of the electric field is only along one direction, the light is said to be **polarized**. An

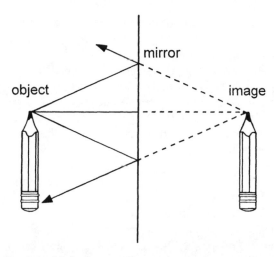

object   image

FIGURE 9-8. The reflection of light from a plane mirror seems to emanate from an image which is the same distance behind the mirror as the object is in front of it. The image is virtual.

analogous situation with a rope would be to pass it through a wooden board with a vertical slit cut in the board. Then the wave that has passed through the slit can vibrate only vertically and not horizontally. There are materials which have similar properties for transmitting or absorbing electromagnetic waves. Commercially available Polaroid filters allow light waves to pass only if the light is polarized along one direction, the polarizing axis of the material. Since light reflected from some surfaces is partially polarized with its electric field parallel to the reflecting surface (usually horizontal), polaroid glasses with their polarizing axes vertical will absorb all of this reflected light. Hence, drivers can remove much glare (reflected sunlight) by wearing polaroid glasses.

Polaroid glasses have also been used in movie theaters to create three-dimensional (3D) images. Two projectors are needed for these 3D movies, one of which utilizes light polarized perpendicular to the ground. The other utilizes light polarized parallel to the ground. If the viewer wears special glasses with one lens having its polaroid axis perpendicular to the ground and the other having its polaroid axis parallel to the ground, the viewer will see an image through the left lens of the glasses which originates from one movie projector and a second image through the right lens of the glasses which originates from the second movie projector. Thus, the viewer is given two different images of the same scene, similar to seeing through two eyes that are spatially separated. These superimposed images appear 3D to the observer.

Holography

Holography is another photographic method used to produce a 3D image. Holographs use **lasers** (light amplification by stimulated emission of radiation). Lasers provide a light source with a single color and a constant phase relationship between its emitted wave fronts. The laser beam is divided into two parts, one of which directly illuminates a film while the other part strikes an object and is then reflected onto the film. The interference between these two portions of the laser beam produces a pattern on the film that contains 3D information about the object. The hologram is used by placing it back into the laser beam. The pattern in the film causes the light passing through it to be deflected so that it appears to come from the original object.

Holography has been used in many science fiction films. For example, in *Star Wars*, Princess Leia transmits her call for assistance via a holographic image that is projected from one of the robots (34 min). This seems technically feasible.

By contrast, the working of the holograph device used in *Total Recall* (96 min) is very difficult to understand. We do not see a laser beam projected from the device, which is worn on the wrist, to the area in which the image appears. Further, the image can be projected completely around solid objects. In view of the Law of Reflection and Snell's Law, it is difficult to imagine how this is possible without a light beam being reflected from surfaces. It should also be noted that a holographic image would not have any substance to it. Thus, in the same scene in *Total Recall*, when a holographic image is surrounded by soldiers who fire into it without shooting one another, we have an impossibility. Their bullets would pass through the image and strike the soldiers standing on the opposite side of the image. In addition, we hear sounds apparently emanating from the position of the image. Sound waves travel quite differently than electromagnetic waves so that the origin of this auditory phenomenon is also inexplicable.

Exercises

1. *Waves:* How many vibrations per second are represented in a radio wave of 120 MHz?

2. What is the frequency of a vibration whose period is 0.05s?

3. *Wave Speed:* What is the wave speed of a water wave whose wave length is 10 m and whose frequency is 3 Hz?

4. *Interference:* Suppose that you stood at a site at which two sound waves were undergoing destructive interference, such that the wave crest of one wave coincided exactly with the trough of another. Suppose also that the intensity of each set of waves was identical? What would you hear?

5. *Doppler Effect:* Suppose that your were driving in a car toward a stationary siren. What would happen to the pitch of the sound as you drove past the siren?

6. *Sound:* Could you hear sounds of 10 Hz? Could you sense these sound waves in some other way? Might your pet hear these sound waves?

7. *Speed of Sound:* What is the speed of sound at 30 °C?

8. Dolphins emit ultrasonic waves with frequencies as high as 250,000 Hz. What would be the wavelength of such a wave in air at 0 °C?

9. *Resonance:* Suppose that you have two identical tuning forks and you set one into motion near the second. What will happen?

10. *Loudness:* What is the increase in sound intensity of a quiet radio above the rustle of leaves?

11. *Shock Waves and the Sonic Boom:* What is the Mach number of a plane traveling at 1,480 miles/h in air at 0 °C?

12. *Light:* If you were spearfishing from above the water would you aim your spear above, directly at, or below the apparent position of the fish?

13. *Polarization:* What percentage of light would be transmitted by two Polaroids pressed together with their polarization axes aligned? With their polarization axes at right angles to each other?

14. *Holography:* Why is it not possible to make a hologram of a 3D object using a source of white light?

BIOLOGY

Chapter Ten
Characteristics of Living Things

Biology is the study of living things, and the work of biologists is to study living things. Biologists need to know which things are living things in order to establish the realm and range of their discipline. All of us humans are interested in understanding what it means to be alive and learning the variety and nature of the other kinds of living things with which we share our planet. Of course there are many things that we encounter daily and consider alive—beginning with ourselves. Our family, all the humans we know, our pets, the houseplants, and a cricket in the basement would all qualify if we began a catalog of familiar living things. We could move outside the home and find birds, squirrels, mosquitoes, spiders, trees, grass, dandelions, mushrooms, and many other living things. Most of us are aware also that there are living things too small to see without the aid of a microscope. Some of them, such as the yeast cells that cause bread dough to rise, are helpful to us; some of them, such as the bacteria that cause pneumonia, produce illnesses; some of them, such as the bacteria living in soil, rarely draw our attention at all.

As long as we consider examples that everyone already knows about, it seems a simple matter to distinguish living things from nonliving things. Just as it was easy to think of examples of living things, it is also easy to think of things that everyone agrees are not alive. Starting inside our homes again, we might think of the furniture, the appliances, the carpets, and so on; outside the home we find the automobile, the sidewalks, a streetlight, and other things that everyone agrees are nonliving.

Instead of thinking about familiar things, imagine now an unfamiliar situation. You are an astronaut who is the mission biologist. It is your job to observe and describe specimens collected from visits to other planets. The first thing you will need to establish is whether a specimen is a living thing, and thus your task to characterize. Specimens are usually collected "in the field," that is, where they normally are located in the natural world. When you encounter a specimen in the field and observe it, there are some characteristics to look for that will help to determine whether it is a good candidate for a living thing.

1. **Movement.** When something moves, apparently of its own volition, it becomes a candidate for a living thing. A boat tied to a dock and a flag blowing in the wind both move. Careful observation reveals that the water currents and the wind gusts are responsible for the movement. Water and wind are not responsible for the movement we see when we watch a fly, a grasshopper, and a fish. However, not all living things are capable of movement; notable in this category are plants. They sink their roots into the ground to get water and minerals and remain to live in that one place. Therefore, while movement can be an indicator that something is alive, some nonliving things move, and some living things do not move.

2. **Growth and Development.** The most familiar kinds of living things are plants and animals, and before we finish grade school we typically have conducted projects demonstrating the growth of plants from seeds or perhaps the formation of a chick as a hen's fertilized egg develops. As children we are actually aware of our own growth and development. Some of us have raised pets in our homes and watched their growth and development. Some things that are able to get larger, that is, to grow, are not alive. Those of us who live in areas that have very cold winters with snowfall have seen the melting snow form icicles along the roofline. As the Sun melts snow and the water runs down over the icicle, it is cooled and forms a new layer of ice that enlarges the icicle. The icicle is not alive, but it does grow. Its development has a rather consistent form in the shape of a cone. Clearly, just because something displays growth and development does not mean it is alive.

3. **Responsiveness.** There are many ways in which we encounter plants and animals reacting to the world around them. If someone hits you, you might respond by hitting back or by moving away to avoid further blows. When you sprinkle food on the water of a fish tank, the fish become more active and dart around snatching the food. A houseplant grows in the direction of the light it needs to survive. It is possible to observe nonliving things involved in what appears to be a response, even though it is not. For example, the increase in length of an icicle might be interpreted as a response in which growth toward the ground occurs.

4. **Complex Organized Structure.** Most of the plants and animals we routinely encounter demonstrate this feature well, beginning with ourselves. It is just as true for a fly or a chrysanthemum plant. Once again, this characteristic by itself is not diagnostic for a living thing. There are many things that have complex organized structures but are not alive. Examples include many mechanical creations of humans, like automobiles, computers, clocks, and televisions.

All of these characteristics could be evaluated by an observer in the field. No single characteristic by itself indicates that a specimen is a living thing. The specimen is a very good candidate for a living thing if it displays two, three, or all of the field characteristics. If it does display multiple field characteristics, it is a good idea to study it further. The best way to do so is to collect a sample for examination in the laboratory. At this stage the investigator needs special training and equipment to evaluate whether the specimen is a living thing. We next examine some characteristics to check out in the laboratory.

1. **Metabolism.** Living things carry out organized groups of chemical reactions that build up their body parts, thus sustaining them and allowing them to grow, or that break down worn out or injured body parts as well as generate energy within the body. Biologists and biochemists have developed various tests to detect these activities of a living thing.

2. **Energy Acquisition.** All living things require an external source of energy to maintain themselves and carry out their activities. The living things we know about use only two sources of energy—the radiant energy of the Sun or the chemical energy of small molecules. Most of the living things with which we are familiar

either capture solar energy by the process of **photosynthesis** or acquire energy-rich food molecules that have been formed as a result of photosynthesis and then release the energy stored in the chemical bonds of these food molecules.

3. **Reproduction.** Living things reproduce their kind, and this is a process which might occasionally be seen in a field study. Many of the living things whose reproduction is easily observed may not reproduce frequently enough to count on seeing it during an observation period. It is sometimes also difficult to get living things to reproduce in the laboratory, especially if they reproduce sexually and require the mating of two types of individuals. This characteristic is often not necessary for the identification of active living things, but it can be very useful for specimens that are collected in a dormant stage and for microscopic organisms. Seeds, spores, or cysts are forms that reveal their living status when they are cultured in the laboratory.

Laboratory observations of these processes bring one closer to the determination that a specimen is a living thing. The reason that they do so is that the processes are characteristic of **cells.** To identify a specimen as a living thing, one must verify that it is composed of a cell or cells. This characteristic is the definitive one. All living things are composed of one or more cells.

An individual living thing is called an **organism** and may range in composition from one cell to trillions of cells. Humans are made of trillions of cells and, therefore, are multicellular organisms. An amoeba living in a pond is composed of only one cell; it is a unicellular organism. There are many different kinds of organisms living on Earth, and each type is a **species.** Humans are a species, and our scientific name is *Homo sapiens.* Corn is a species of plant, and its scientific name is *Zea mays.* The yeast we use to make bread rise is another species; its scientific name is *Saccharomyces cerevisiae.* By using a scientific name, scientists all over the world can discuss the same organism unambiguously, no matter what their language.

The Cell Theory

By the middle of the 19th century, scientists using microscopes to study cells realized that everything they regarded as a living thing was turning out to have cellular structure. They came to recognize three ideas that we now call **the cell theory.** First, the cell is the basic unit of structure and function in living things. Second, all living things are made of one or more cells. Third, all cells arise from preexisting cells.

Cell Structure and Function

Since all living things are made of cells, it is important to learn some of the characteristics of cells. All cells have a structure called the **cell membrane** that encloses the internal contents called the **cytoplasm.** Also present inside the cell is **genetic material** which contains the encoded information that allows a cell to reproduce itself. In some cells the genetic material is free in the cytoplasm; such cells are called **prokaryotic** cells. Other cells have their genetic material located within a double membrane to form a structure in the cytoplasm called the **nucleus**; such cells are called **eukaryotic** cells. The cells of bacteria, for example, the bacterium *Escherischia coli* which lives in our intestines, are prokaryotic cells. The cells that form a human body are eukaryotic cells.

Our own body cells are very small, and they, like most kinds of cells, can be seen only with a microscope. Our knowledge of cell structure has been aided greatly by using electron microscopes to study cells. Instead of putting light on a specimen, they use beams of electrons, and the lenses are magnetic rather than glass. Specimens are permeated with a tough plastic material that can maintain the specimen's arrangement and withstand the electron beam. Then the specimen is sliced very thinly using specially made knives of glass or diamond. When the slices are examined in the electron microscope, we see that there are many structures in the cytoplasm of cells. They reminded biologists of the organs that make up human bodies and carry out specific functions. Accordingly, they named these tiny structures **organelles.** Their distribution and function have been studied extensively.

To determine the functions of organelles requires studies that go beyond simply examining the microscopic appearance of the organelles. Biologists have devised many clever, often difficult, expensive, and time-consuming procedures to conduct such studies. The results from decades of research by many researchers allow us now to correlate cell structures with their functions.

Selected Organelles and Their Functions

Cell Membrane. This structure is also called the plasma membrane. It serves as the boundary which separates the cell's contents from the extracellular environment. This function is due chiefly to the lipid molecules that are distributed as a bilayer. Because of the lipid character, even though this part of the membrane is very thin, it acts as an effective boundary between the cytoplasm and the extracellular environment, which are both watery environments. Fatty materials and watery ones do not mix. Think of fat droplets floating on the top of soup to remind yourself of the incompatibility of watery and fatty substances. The cell membrane also determines which things can cross it, either to enter or leave the cell. Because the cell membrane allows some molecules to cross and others are barred, it is said to be **selectively permeable** or **differentially permeable.** Protein molecules that are incorporated into the lipid bilayer are chiefly responsible for this function. Some of them form channels that can open and close. Some of them bind certain molecules and "escort" them across the membrane. Some are the means by which cells use some of their energy to bring in molecules that are important for the cell to have in adequate amounts and which may be available in limited supply.

Protein molecules also act as receptor molecules on the outside of the cell membrane. These molecules detect the presence of external molecules and then signal the inside of the cell to respond. For example, when someone has hay fever or a cold, molecules called histamine are released. The cells of the nose that produce mucus have receptors that are able to detect these molecules. The presence of histamine causes the mucus-producing cells to secrete large amounts of very thin, watery mucus. This response helps wash out of the nose the offending stimulus. In the case of hay fever, there is apparently no danger to the body, and the response seems unnecessary. In the case of a cold, a virus is invading the body, and the response helps to rid the body of an agent that could cause harm to body cells.

Cytoplasm. Historically cytoplasm meant the material within the cell membrane. Plant and animal cells were seen as having a nucleus surrounded by cytoplasm, while bacterial cells were seen to lack the nucleus. When very high magnification of cells became possible with the electron microscope, it was soon apparent that there were many structures contained within the region called the cytoplasm. Currently the term cytoplasm is sometimes used in the original sense and sometimes modified to mean the material in which the organelles are found. Sometimes the term **cytosol** is used to convey the latter sense unambiguously. A major role of the cytosol is to begin the release of energy from glucose molecules. It contains the enzymes of the metabolic pathway called glycolysis.

Ribosomes. Ribosomes are tiny structures of the cytoplasm. They are made up of proteins and ribonucleic acid (RNA) molecules. Their extremely important function in the cell is the synthesis of proteins. Ribosomes have the ability to "read" from molecules called messenger RNAs the coded information that specifies the sequence in which amino acids must be assembled to form the various protein molecules of the cell. Ribosomes are found in both prokaryotic and eukaryotic cells.

Mitochondria. Mitochondria are organalles that, in many organisms, have a shape similar to a kidney bean. When they are sliced open, their interior is quite different from a kidney bean. It is made up of folded membranes. Mitochondria have the task of providing energy to the cell in a form that it can have available for prompt use anywhere energy is needed. Because of this role mitochondria are described as the power plants of the cell. Mitochondria are found in eukaryotic cells, and they are never present in prokaryotic cells.

Chloroplasts. Chloroplasts are organelles that carry out photosynthesis, the metabolic process that traps the energy of the Sun and stores it in energy-rich food molecules. The green pigment chlorophyll is essential to this process and gives many photosynthetic organisms a green color. Photosynthetic organisms that are not green still have the green pigment, but it is masked by other pigments. Like mitochondria, chloroplasts are found only in eukaryotic cells and never in prokaryotic cells. A further similarity to mitochondria is their internal structure of membranes, although the arrangement of the membranes is different from that of the mitochondrion. The chloroplast's shape is typically more like a lens than a kidney bean.

Endoplasmic Reticulum and Golgi Complex. These structures are made of

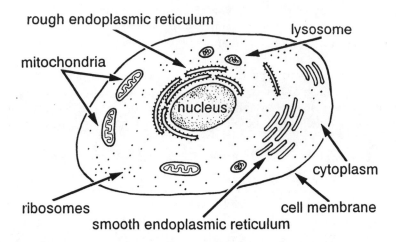

FIGURE 10-1. A typical animal cell. This drawing shows a cell visualized by transmission electron microscopy. The specimen is a very thin slice of a cell. Biologists had to look at many such photographs to construct three dimensional versions of the appearance of organelles. The task is similar to looking at the slices of a cucumber. It is possible to look at each slice and figure out what the whole cucumber looks like.

membranes folded in characteristic patterns. The endoplasmic reticulum (ER) ranges from broad stacks of membranes to more rounded, vesicular profiles, while the Golgi complex is formed from narrower, tapered stacks of membranes. These differences in appearance signal different functions also. The ER is found in two forms: rough ER that is associated with ribosomes and smooth ER that has no ribosomes. The smooth ER has groupings of membrane proteins that carry out a variety of metabolic reactions. Among them are the biosynthesis of cholesterol and other lipids. Even though most people probably think of cholesterol as "bad," it is actually a very important constituent of animal membranes. While excess cholesterol is certainly associated with health problems for humans, no cholesterol would be a serious problem also.

Smooth ER receives proteins into the interior spaces of the membrane system as they are synthesized on the ribosomes. The proteins then have various types of sugars attached to them. Little packets of protein product, called transport vesicles, bud off the rough ER. They migrate to the Golgi complex and merge with its membranes. Here additional attachment of sugar molecules takes place. The finished products have a variety of fates, including the formation of lysosomes, microbodies, and storage vesicles.

Lysosomes. Lysosomes are small, spherical bodies which contain a powerful array of proteins called enzymes that are able to digest most of the types of molecules that exist in living things. Their enzymes called proteases digest protein, those called lipases digest lipids, those called nucleases digest nucleic acids, and those called glycosidases digest carbohydrates. One of the mysteries surrounding lysosomes is why they do not digest themselves. When lysosomes are newly formed, they look like a membrane sac with a clear interior. When they are in the process of digesting something, they look swollen, sometimes misshapen, and their interior is no longer clear, but filled with partially digested material. These organelles serve as the cell's digestive system when they break down things brought into the cell from the outside. When an amoeba engulfs a food particle, lysosomes then enable its digestion. Lysosomes also act as cellular "recycling centers". When a cell part is worn out or injured, it will be digested by a lysosome. The broken down molecules can then be reused by the cell. Some fascinating photographs made with electron microscopes show lysosomes digesting mitochondria.

Lysosomes are found in eukaryotic organisms, but not in prokaryotic organisms. In plants they often exist in a modified form.

Vacuoles. Vacuoles are not really vacant, but they may look empty when we view them because their contents are often colorless, transparent fluids. Sometimes the contents are removed during specimen preparation, and then the vacuole is vacant, but artificially so. Two interesting and important kinds of vacuoles are the central vacuole of plant cells and the contractile vacuole of various protists, such as an amoeba. The central vacuole may be the largest organelle of some plant cells. As long as water is available in sufficient quantities, the central vacuole remains full and plump. This contributes to the support of the plant body. When the central vacuoles are not full of fluid, the plant wilts.

The contractile vacuole of an amoeba allows it to live in ponds where water constantly rushes into the cell. The contractile vacuole collects the water and squirts it out of the cell. Thus the proper amount of internal water is maintained, and the cell can function properly.

Cytoskeleton. The cell's shape is determined by the configuration of the rod-like elements of the cytoskeleton. They are of different sizes and made of different kinds of molecules as well. They are functionally quite different from familiar kinds of animal skeletons in that they are dynamic structures. They change their arrangement readily. The cytoskeleton of a cell can be seen to have one kind of configuration when the cell is at rest, and when it begins to move, the cytoskeleton is transformed into another arrangement. Some of the elements of the cytoskeleton are microtubules, microfilaments, and intermediate filaments. Further, the structures found in cytoplasm do not simply drift in the cytosol. They are attached to elements of the cytoskeleton.

Nucleus. The nucleus of a eukaryotic cell contains the genetic material that is responsible for the cell's ability to divide to form new cells with the characteristics of a given type of organism. This genetic material that is responsible for heredity in both eukaryotic and prokaryotic cells is a type of nucleic acid called **deoxyribonucleic acid** or DNA. The DNA of the nucleus is closely associated with a variety of protein molecules and a small amount of the other type of nucleic acid called RNA. These components form **chromatin** which is sometimes dispersed loosely in

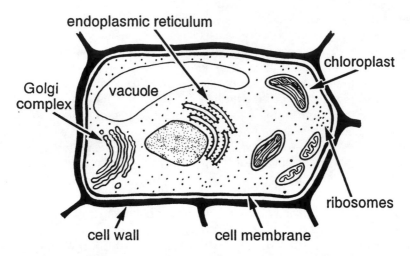

FIGURE 10-2. A typical plant cell. This drawing shows a plant cell viewed by transmission electron microscopy. Which organelles do you see that are found in both plants and animals? Which structures are found only in plants?

the nucleus and sometimes condensed into **chromosomes.** In most organisms chromosomes are seen when the cell is about to divide.

Cell Wall. Cell walls are found in both prokaryotes and eukaryotes. Cell walls are located outside of the cell membrane. They protect the fragile cell within it, provide structural support for the organism, and permit the cell to be placed in a wide variety of watery environments without danger of bursting. Cell walls are made of a variety of polymers. Interestingly, the molecules that form them are dramatically different among the kingdoms which are discussed below. A polymeric material called peptidoglycan is the basic material of prokaryotic cell walls. In plants and protistans the major component of the cell walls is cellulose. Fungal cells walls are built with polymers called glycans.

Our kingdom (see below), Animalia, is the only one in which cell walls are not found. Since animal cells have no cell walls, while bacterial cells do, this difference between the cells is what permits us humans to use the antibiotic penicillin. An **antibiotic** is an agent that kills or limits the growth of some kinds of living things. We use the antibiotic penicillin to help our bodies overcome bacterial infections. It works by interfering with the ability of bacterial cells to form their cell walls. Thus, when we swallow a penicillin tablet and the chemical is taken up in the bloodstream to be distributed throughout our bodies, the penicillin does not interfere with our cells because they do not form cell walls. In contrast, the bacterial cells need their cell walls in order to survive and function properly.

Whenever we can identify a difference in the way our body cells operate compared to organisms that cause medical problems, there is an opportunity to control the organism when it causes illness. Another antibiotic used to protect humans from bacteria is tetracycline. It is thought to work by inhibiting protein synthesis. Since prokaryotes and eukaryotes have different types of ribosomes, this chemical is useful because it works on bacterial ribosomes but not our eukaryotic ribosomes.

Flagella and Cilia. Flagella are whiplike appendages used by cells to propel them from place to place. They are found on both prokaryotic and eukaryotic cells, although they are made of completely different molecules. Examples of cells that have flagella are sperm cells, the bacterium *E. coli*, and the intestinal protistan parasite *Giardia*.

Some cells have numerous short projections either all over their surfaces, or in localized clusters. These are cilia. When they beat, they cause the cell to move if it is a free cell. Among the ciliated organisms that are common in ponds and lakes are species of *Paramecium*. When one tries to look at the organisms with a microscope, they just zip by and are very difficult to observe. Adding special preparations to slow them down allows them to be observed. Sometimes cilia are found on cells that are anchored in place with a group of cells. This is the case for cells that form the walls of the airways in our lungs. Their beating moves mucus out of the airways and up to the throat to be swallowed.

Table 10.1 lists these important cell structures and their distribution. These are not the only cell structures that are known, but a representative set of important, widely distributed structures. Not only are there more that are found in "typical" cells, but also there are many specialized structures that are found only in certain organisms or in certain types of cells within organisms.

Cell Size

Most cells are very small and can be seen only with the aid of a microscope. Many kinds of cells fall in the range from 1 to 100 μm. (See Chap. 1 in the Physics section for a discussion of metric measurement.) Some cells are much larger. For example, the egg cells of many animals are large enough to see. You have certainly seen the yolk of a hen's egg; it is the cellular part of the egg. Perhaps you have seen a cluster of frog's eggs in a pond. The cells of the green alga *Acetabularia* are 2 to 3 cm long. The long extensions of neurons, the impulse-carrying cells of the nervous system, may be 1 m or longer. These cells are exceptional. Cells are usually very small.

An important factor that dictates a small cell size is the relationship between a cell's volume and its surface area. If we think of a cell as a sphere, its surface area increases with the square of the radius. In contrast, its volume increases with the cube of the radius. That means that when a cell doubles its radius, there is eight times more volume but only four times more surface area. Since cells must receive nutrients and get rid of wastes for their entire volume by way of their available surface area, a cell that is very large probably cannot have adequate exchanges take place to provide for its needs. Also, since cell activities are often triggered by the reception of a signal at the surface of the cell, the response time of the cell

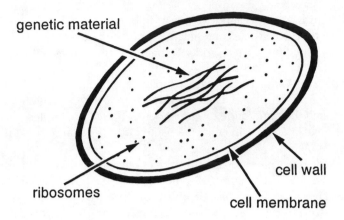

FIGURE 10-3. A representative bacterial cell. This drawing illustrates the appearance of a bacterial cell viewed by transmission electron microscopy. Notice how simple it is in comparison to the plant and animal cells.

depends on the cell's ability to transmit the signal to its interior and initiate an appropriate response. If the distances are too great, the cell will be too slow to respond as the greater distances to interior structures mean that more time is needed to receive the signal and to carry out the response over greater intracellular distances.

Organisms

An individual living thing is called an **organism.** Thus you are an organism, a sparrow is an organism, a redwood tree is an organism, and so forth. Organisms like these can be seen easily and are called **multicellular organisms** because they are made up of more than one cell. Many kinds of organisms contain only one cell; they are called **unicellular organisms.** These are too small to see with our eyes, but they can be seen with both light microscopes and electron microscopes. Recall that a kind of organism is called a species. Estimates of the number of different species on Earth vary somewhat, but the lowest estimates are that the Earth is home to over 2 million different kinds of species.

Kingdoms of Organisms

With so many different kinds of organisms living on Earth, it would be very difficult or impossible for biologists to deal with all of these living things if they did not organize their thinking and categorize species according to their similarities. Biologists have done this, and the largest groups of organisms they have established are called the **kingdoms.** Not all biologists agree on the characteristics that should be used to establish the kingdoms, and as a result several schemes have

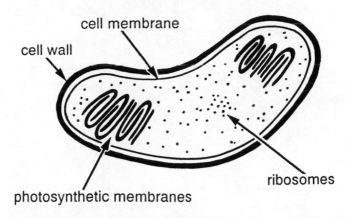

cell membrane

cell wall

ribosomes

photosynthetic membranes

FIGURE 10-4. A representative photosynthetic bacterial cell. In this drawing of a transmission electron micrograph of a photosynthetic bacterium, notice the membranes that give the cell its ability to conduct photosynthesis.

been proposed for sorting species into kingdoms. The most widely used system distributes living things into five kingdoms: Monera, Protista, Fungi, Plantae, and Animalia.

The kingdom Monera is the kingdom of the bacteria. Its members include tiny organisms that live in many different kinds of environments. Some live on and in other organisms, some live in the oceans and seas, some live in the lakes and

TABLE 10.1.

Component	Kingdoms				
	Monera	Protista	Plantae	Fungi	Animalia
Nucleus		x	x	x	x
Ribosomes	x	x	x	x	x
Endoplasmic reticulum		x	x	x	x
Golgi complex		x	x	x	x
Mitochondria		x	x	x	x
Chloroplasts		s	x		
Vacuoles		x	x		s
Lysosomes		x	x		x
Cytoskeleton		x	x	x	x
Cell wall	x	s	x	x	
Flagella/Cilia	x	x	r		x

x = present in all or nearly all members of the kingdom.
s = present in some kingdom members.
r = rare in the kingdom.

TABLE 10.2.

Kingdoms

	Monera	Protista	Plantae	Fungi	Animalia
Prokaryotic or eukaryotic	P	E	E	E	E
Unicellular or multicellular	U/M	U	M	M	M
Autotrophic or heterotrophic*	A/H	A/H	A	H	H
Filamentous body				Yes	No
Cell wall				Yes	No

*An autotrophic organism makes it own energy-rich food molecules from simpler molecules that are readily available as raw materials from the environment. Heterotrophic organisms depend upon autotrophic organisms as their source of energy-rich food molecules. The Greek roots of these words translate as "self-feeders" vs "fed by others."

rivers, and some live in the soil. Some can survive in a wide variety of living conditions, while others die quickly if strictly required conditions are not maintained.

The kingdom Protista includes very small organisms also. They, like the organisms in the kingdom Monera, are also a diverse group. You may have encountered some of them in biology class. Students frequently examine amoebae or *Paramecium* as examples of cells and also to practice using a microscope. Perhaps you have done so. Protists also include parasitic organisms such as the organism that causes malaria and photosynthetic organisms such as diatoms that are found in large numbers in both fresh and salt water.

Some representative organisms of the kingdom Fungi are mushrooms and molds. Many of them are free-living organisms that help to recycle dead material by decomposing it. Some cause diseases in plants and animals, e.g., the rusts that can be devastating to a farmer's crop. Some fungi are beneficial. The mold *Penicillium* is the source of the antibiotic penicillin that has saved many lives. Yeast used to leaven bread is a member of the kingdom Fungi.

The kingdoms Animalia and Plantae are the two whose members are most familiar to us. We are members of the kingdom Animalia, as are worms, moths, birds, jellyfish, and sponges. Pine trees, grass, thistles, ferns, mosses, red bud trees, and century plants are all members of the kingdom Plantae.

These groupings of organisms have been established after biologists gave a great deal of thoughtful effort toward identifying some fundamental characteristics of organisms. To begin to determine the kingdom to which an organism belongs, one determines three fundamental characteristics:

1. Does the organism have cells that are prokaryotic or eukaryotic?
2. Is the organism unicellular or multicellular?

3. Does the organism acquire energy by photosynthesis or (autotrophism) from energy-rich food molecules (heterotrophism)?

Table 10.2 lists the characteristics of the kingdoms. Notice that the answers to the three questions mentioned above yield a different combination of characteristics for each kingdom except for the kingdoms Fungi and Animalia. These two kingdoms are easily distinguished on the basis of their respective characteristic body forms and whether they secrete a cell wall outside of the cell membrane. Fungi have filamentous bodies, and cell walls are present. Animals have neither filamentous bodies nor cell walls.

The Andromeda Strain

Now that you have knowledge of the characteristics of living things, their cellular nature, and the sorting of organisms into kingdoms, you are ready to examine the film, *The Andromeda Strain*. In this film a team consisting of three scientists and a medical doctor work in a secret underground laboratory on a project called Wildfire and try to identify the cause of mysterious mass deaths occurring in the small town of Piedmont, New Mexico, where a satellite fell from orbit. They learn that the satellite was launched to try to collect extraterrestrial lifeforms that might be useful in biological warfare. This endeavor was called Project Scoop, and, indeed, the satellite was designed like a scoop. Apparently, in crashing out of control to Earth, the satellite leaked out something lethal. It was the job of the Wildfire team to detect and characterize "the organism" and to devise a way to control it.

The greenish speck of material they find on the mesh of the satellite's scoop gets larger before their eyes as they examine it microscopically. The team members seem completely willing to regard the specimen as "alive," and it is named the Andromeda strain. They find only one characteristic of a living thing, growth, and everything else they find out about it points to a structure that is not cellular. They characterize it as similar to plastic and determine that it has neither amino acids, which are the building blocks of proteins, nor nucleic acids. Transmission studies employing filters reveal that particles between 1 and 2 μm in diameter are quickly lethal when transmitted by air to a healthy animal.

The size of the deadly particles is consistent with their being small cells. There is no indication of any cellular character, however. The chemical makeup is totally unlike the familiar cells of Earth's organisms, for no cell exists that is devoid of amino acids and protein. Even viruses, which are not cells, contain amino acids and nucleic acids.

Viruses bear some similarity to the unknown specimen. They are often considerably smaller than this specimen, but some share with it a crystalline structure, which is revealed in the electron microscope as particles that have geometric shapes. Examination with the electron microscope prompts the growth and reproduction of the strange material, even though it is in a vacuum and being bombarded with a high voltage electron beam. These characteristics are unknown in any organism on Earth, and they are unknown among viruses.

Viruses are very compact structures. They are made up of molecules that pack very efficiently into a particle that is capable of invading a cell and using its systems to reproduce viral molecules, which in turn assemble to form new viral particles. Biologists generally do not regard viruses as organisms because they are not cells and they cannot give rise to new viruses from preexisting ones without the aid of cells. The Andromeda strain has a crystalline structure like a virus, but unlike a virus, it is able to reproduce itself.

If some of the functional characteristics of cells were present in the Andromeda strain, perhaps we would have to conclude that the unearthly thing is functionally a cell even though it lacks the recognized cellular anatomy of an Earth cell. Studies to detect and characterize metabolism were apparently not conducted. The growth studies that are mentioned seem to be routine bacterial studies that vary such things as the nutrients that supply energy to the cells. These are difficult to understand in light of the other finding that Andromeda uses high energy sources directly to provide for its energy needs.

Overall, the film provides little evidence that the Andromeda strain is a living thing. The fact that the film's scientists do believe it to be living, however, brings up the intriguing possibility that living things can survive in space. Most scientists believe that extreme conditions of temperature and pressure coupled with exposure to radiation make space an environment in which life cannot survive.

The Andromeda strain reminds us that the specimens we encounter may not always be easy to place in discreet categories. In the natural world the levels of organization (see Chap. 13) move from molecules to molecular assemblages. The organelles of living things are molecular assemblages, which in turn are found in the functional groupings that constitute cells. Organelles are not the only molecular assemblages that are found in nature, for viruses are molecular assemblages. Some biologists favor including viruses among living things because they have concluded that the deciding factor should be not cellular structure, but the use of nucleic acid as a genetic code. Viruses do use nucleic acids for their genetic code. They differ from cells in that cells use DNA for their repository of genetic information, while viruses, as a group, use both DNA and RNA. This distinction is a way of dividing viruses into two large groups—the DNA viruses and the RNA viruses. This criterion of using a nucleic acid for a genetic code places the emphasis in defining life on the process of information storage rather than on reproduction. Viruses do have the ability to produce offspring of their own kind, just as cells do. They are unable to do so without enlisting the services of cells. To some biologists this dependence on cells disqualifies viruses from the roster of living things.

A few scientists believe not only that life can survive in space, but also that the first living things on Earth came here from space. This idea is called **panspermia.** Some scientists have proposed that the origin of life on Earth was not an evolutionary process that formed cells from nonliving materials that became more organized and more complex, but instead the first life on Earth came falling into the atmosphere from space.

EXERCISES

1. Compile a table of the observations made about the Andromeda strain in the film. Then determine whether the observation is useful in determining whether Andromeda is living or not. Finally, note how the observation was utilized in the film and whether you agree. Which observations came directly from the work of the Wildfire scientific team? Which do you attribute to other observers?

2. Some cells are able to use the energy of sunlight, through the process of photosynthesis, to capture energy for cellular use. The energy used by Andromedas is more potent than sunlight and would injure or kill known cell types. Look up the electromagnetic spectrum and the kind of damage done to cells by high energy sources such as gamma rays and ultraviolet light. What cell components are damaged? Is it plausible that Andromeda could use, rather than be harmed by, high energy bombardment? Are viruses susceptible to damage by high energy? If so, what damage can you learn of? If not, why are they immune?

3. Particles called viroids and prions that are simpler than viruses have been reported. Viroids are made only of nucleic acid, while prions are made only of protein. Both cause disease. Does their existence make the existence of Andromeda plausible? Why do you think so?

4. When astronauts take trips into space, great care is taken to ensure that they do not bring new organisms to Earth. Why do you think such precautions are taken? When someone discovers a meteorite, is it safe to touch it or take it home?

Chapter Eleven
Cellular Reproduction

Individual cells do not live forever. Therefore, in order to maintain species, whether they are unicellular or multicellular, it is necessary for cells to reproduce. The simplest mode of reproduction is called **binary fission.** In this process a cell divides into two cells, which are usually called the "daughter" cells. Bacteria divide in this way. They have one copy of their genetic material in each cell, and it is duplicated prior to cell division. Then the original cell is partitioned by the formation of new cell membrane between the two copies of the cell's genetic information. After the new sections of membrane are completed, a new cell wall is deposited to complete the binary fission.

When eukaryotic cells divide, the process is somewhat more elaborate, but the basic idea is the same: what began as a single cell divides into two daughter cells. In achieving this end, the cell carries out the division of the nucleus, a process called **mitosis,** and then the division of the cytoplasm, which is called **cytokinesis.**

In contrast to bacterial cells in which the genetic information is present as a single circular molecule of DNA, eukaryotic cells have more than one molecule of DNA. Each molecule of DNA is associated with special proteins that permit it to exist either in a loosely packed state or in a very condensed state. It is the condensed form that is usually referred to as a **chromosome,** although technically the DNA also forms a chromosome when it is in its loosely packed, nonmitotic form. When mitosis occurs, the DNA condenses into discreet chromosomes after it has been duplicated. Then the two sets of chromosomes separate. Cytoplasmic structures called **microtubules** form the **spindle,** whose role is to pull apart the two sets of chromosomes.

About the time that the chromosomes condense, the nuclear membrane seems to disappear as it breaks up into tiny vesicles to allow space for the movement of the chromosomes and the eventual formation of two nuclei, rather than one. After the chromosomes have separated, the nuclear membrane reforms as two complete nuclear membranes, one around each set of chromosomes. The chromosomes lose their dense packing and unravel to take on the usual appearance of a nondividing cell, which is called an interphase cell. In this state no separate chromosomes are seen, but the association of DNA with protein is referred to collectively as the **chromatin** of the nucleus.

Cell division is not complete until the cytoplasm is divided into two parts also. This process has the appearance of two drawstring bags being tightened in mirror image fashion. Tiny contractile filaments in the cytoplasm, the **microfilaments,** are crucial in this process. It is their contraction that is responsible for the constriction of the cytoplasm that completes the formation of the two daughter cells.

Diploid and Haploid Cells

The cells of most eukaryotic organisms have a double set of genetic information. When their chromatin condenses into chromosomes, they are present as pairs. This condition makes a cell **diploid.** (Notice the first part of the word is di-, which means two. There are two sets of the cell's genetic material.) Most of the cells of your body are diploid. When they divide, they are seen to have 46 chromosomes altogether, or 23 pairs of chromosomes.

Prokaryotic cells have only one copy of their genetic material. They are called **haploid** cells. Their basic life cycle is very simple: haploid cells divide by binary fission to give rise to new haploid cells. Some eukaryotic organisms, however, have a life cycle in which they spend time in a diploid form and in a haploid form. Ferns are organisms that have this type of life cycle. The form that we normally think of first, with its lush green fronds that can form dense foliage in forests and gardens, is the diploid form. The haploid form is very small and easy to overlook. Fern generations alternate between the diploid and haploid forms.

Fertilization and Meiosis

When the life cycle of an organism includes both diploid and haploid stages, there must be a means of switching from haploid to diploid and diploid to haploid. The former switch is accomplished by the process of **fertilization** and the latter by **meiosis.** When fertilization occurs in humans, two haploid cells fuse to form one diploid cell. The haploid cells are the gametes or sex cells produced by the ovaries and testes. These are the only cells in a humans's body that are haploid. When an egg that has been produced by a woman's ovary fuses with a spermatozoon that has been produced by a man's testis, a zygote results that will develop into an embryo that eventually will be born as a baby boy or girl.

The process of meiosis is sometimes called reduction division because it is the type of cell division that permits cells to shift from diploid status to haploid status. Meiosis is the process that occurs in the ovaries and testes of humans to form haploid gametes. It differs from the mitotic type of cell division not only because it produces products that are not the same as the starting cell, but also because the process includes two rounds of cell division rather than one. Just as in mitosis, the cell prepares for division by duplicating the DNA. Since the starting cell is diploid, that means there are four copies of the cell's genetic information at this point.

In the first round of meiotic cell division, the pairs of chromosomes are separated with the two resulting daughter cells having one chromosome from each pair. Since the DNA was duplicated before meiosis began, each of these cells contains half of the diploid set of chromosomes, but each is present in duplicate. The second round of meiotic cell division distributes one copy of each chromosome to a daughter cell, which is haploid. When meiosis is complete, what began as one diploid cell has divided twice to form four cells, which are haploid. In the testis, all four cells normally mature to form functional spermatozoa. In contrast, the ovarian process yields only one functional cell, the egg or ovum.

Organisms that use meiosis and fertilization carry out **sexual reproduction,** and their offspring have a genetic contribution from two parents. Organisms that keep their genetic status the same, either haploid or diploid, are using **asexual reproduction.** Some organisms are capable of reproducing in both ways. For example, you may have seen some kinds of grass produce "runners" along the ground that can take root and form a new plant. Such reproduction is asexual. Those same grass plants can also form flowers for sexual reproduction which will produce seeds.

Development

When multicellular organisms reproduce sexually, the new individual begins life as a single cell. That cell must multiply and give rise to the characteristic body features of the species and all of its specialized types of cells. Altogether these processes constitute the **development** of the organism. The process of establishing the characteristic body plan is called **morphogenesis,** and the formation of specialized cells is called **differentiation.** When you began your development, you were a single cell called a **zygote,** and you had no recognizable human features. By the time you were born, indeed even before then, your basic human body plan could be recognized. Part of the genetic information of the DNA in human cells is responsible for setting in motion the events that lead cells to multiply and migrate in the ways that form a human body plan.

Biologists sometimes describe the genetic information of a species as being divisible into two sets. The first set is the genes that code for the development of an organism, as described above. The second set is the "housekeeping" genes that are responsible for the structural molecules of the cells and the molecules that enable the cells to conduct their day-to-day cellular activities that collectively make up the life of the organism.

Genes

A gene is a portion of a DNA molecule that codes for a functional product which is most often a protein, but may also be a molecule of RNA. Diploid cells have two copies of each gene. However, the two copies may not be identical, because many genes have alternative versions, called **alleles.** If the two alleles are identical, the cell is called **homozygous** for that trait, but if the two alleles are different, the cell is called **heterozygous.**

In some cases biologists have learned the activity controlled by particular genes. Examples of such genes are the ones that determine a person's blood type. Humans have blood that falls into one of four categories: A, B, AB, or O. These categories are determined by the character of a glycolipid molecule found in the membranes of red blood cells. The "glyco-" or carbohydrate portion of the mol-

TABLE 11.1. Genetics of red blood cells. The gene that controls the membrane glycolipid which gives red blood cells their primary blood type character is called I. It has three alleles called A, B, and O. They code for enzymes that produce variations in the carbohydrate portion of the membrane glycolipid, as depicted.

Phenotype	Genotype(s)	Membrane glycolipid	Red blood cell
O	OO	Lipid-sugars	OH
A	AA,AO	Lipid-sugars-NGal	OH-NGal
B	BB,BO	Lipid-sugars-Gal	OH-Gal
AB	AB	Lipid-sugars-NGal	OH-NGal
		Lipid-sugars-Gal	OH-Gal

The genotype may also be expressed by combining the allele symbols with the gene symbol. I means the allele is dominant, and i means the allele is recessive. Then AB blood is symbolized by I^{AB}, O type blood by ii, A type blood by $I^A I^A$ and $I^A i$, and B type blood by $I^B I^B$ and $I^B i$.

ecule is the determining part. The basic molecule that is formed by the red cells of all humans is called the H substance. A person whose red cells have H substance as the final product belongs to the O blood type. Such a person does not have the protein molecules called enzymes to proceed beyond the H substance. Other people have the allele that produces an enzyme that adds to the H substance the sugar called N-acetylgalactosamine. These people belong to the blood group A. Some people have the allele that produces an enzyme that adds the sugar galactose to the H substance. These people belong to the blood group B. Yet another group of people have both of these enzymes formed because their diploid cells have the gene for A character on one chromosome of a pair and the gene for B character on the other chromosome of the pair. Then the red cells have both A and B character, and such people belong to the blood group AB.[1]

The observed character that is transmitted genetically is called the **phenotype.** In the example of the red blood cells' membrane molecules described above, the four blood groups also constitute the four phenotypes for this character. However, an individual also has a **genotype** for a character. The genotype is the pair of alleles present for that character. A red blood cell could have the A phenotype in two ways: both alleles could be A or one A allele could be paired with an O allele. The first situation is homozygous, while the second is heterozygous. The same possibilities exist for the B phenotype. The AB phenotype can exist only heterozygously, and the O phenotype can exist only homozygously. (See Table 11.1.)

Some genes can be identified by observing the inheritance of the traits they control, but biologists do not yet know the precise cellular action controlled by the gene. There are many such traits known. An example is the Darwin's point, also called the Darwin's tubercle. It is a small bump that appears on the upper rim of

---

[1] The red blood cells of humans have no nuclei; however, they develop in the bone marrow from cells that do have nuclei. As red cells mature, they lose their nuclei and other familiar organelles to become specialized for their role in gas transport.

the external ear in some people. The gene that controls it somehow dictates a thickening of the cartilage of the ear at this point. This is a feature of morphogenesis, albeit a rather minor one.

The Darwin's point is a dominant trait. This means that if one allele has the gene for this trait, it will be expressed. The absence of a Darwin's point, which produces an even outer edge to the external ear, is recessive. It is necessary for both alleles to code for this type of ear development in order for it to be expressed. If the dominant allele for a Darwin's point is present with an allele for a smooth edge, the Darwin's point is produced. The allele for the Darwin's point has dominated over the presence of the recessive allele for a smooth ear edge. Dominant alleles are usually symbolized by capital letters and recessive alleles by small (lower case) letters. In Table 11.1 the gene I, which codes for ABO character on red blood cells, stands for a dominant allele. Since there are two dominant alleles, A and B, a superscript is used to indicate which dominant allele is present. The recessive gene i stands for the allele that produces O character.

Mutations and Them!

Even though there are many small differences among individuals of a species, on the whole they are more similar than dissimilar. Humans are more like one another than like other primates such as lemurs, gorillas, chimpanzees, and orangutans. Even though the distribution of different alleles throughout the human populations produces genetic variability, it occurs on a relatively small scale in comparison to the changes that are depicted for an ant species in the film *Them!*. Mysterious sounds, large-scale destruction of property, and the death of a storekeeper confuse officials at first, but after enlisting the aid of biologists who specialize in the study of ants, they determine that giant ants are at large in the desert. The ants appear to be a species of local desert ants that have somehow become giants. The baffled biologists can only speculate why the ants have become about 10-feet long. Their educated guess is that atomic bomb testing in the desert years earlier produced **mutations** in some local ants.

A mutation is a change in an organism's DNA. If such a change occurs in the body cells of a multicellular organism, it dies out with the individual. The body cells are called somatic cells, in contrast to the sex cells or gametes. Therefore, this type of mutation is called a **somatic mutation.** If, however, such a change occurs in the DNA of the gametes, it may be transmitted to offspring, thus producing a new allele in the species. This type of mutation is called a **germinal mutation.**

Radiation is one physical agent that causes mutations. (See the physics section for a discussion of radiation.) When radiation strikes DNA, it can alter the DNA's chemical character, which is essential to the proper "reading" of the genetic code, because the shapes of the DNA's structural subunits, called nucleotides, establish their functional identity. If radiation striking DNA alters a nucleotide, its new shape may be interpreted as some other nucleotide and thus misread; it may even be so drastically changed that the genetic code at that point becomes unreadable. In the first instance the product for which the gene codes may be nonfunctional because it is incorrectly composed, while the second situation produces no product at all.

In *Them!* not only were the ants giants, but also their development was altered. They hatched from eggs as adults rather than going through larval and pupal stages. On the one hand, it is difficult to imagine this drastic alteration in development occurring by way of a single mutation. On the other hand, it is improbable that a series of mutations could be brought about in already-mutated individuals to effect multiple changes in the genes controlling development. For example, if we consider a high probability for the first mutation of 1 in 1,000, and then the same probability for two other mutations that must take place in succession to alter the development, the final probability of achieving these three mutations in one individual is 1 in 1 billion. Then if the mutations are recessive, there must be more than one individual who carries them, and these ants must breed appropriately to have the new recessive phenotype expressed.

On the whole, then, it is hard to conclude that this change in the development of the ants is plausible in the short amount of time that had elapsed since the atmospheric nuclear testing. It is much more likely that a mutation in the genes that operate the morphogenesis of the ants would fail to develop a viable ant and that the mutation would be the type called a **lethal mutation.** Many such mutations in development are known. It appears to be more likely for significant developmental mutations to result in the death of the developing individual than for them to produce altered development of a viable individual.

Interestingly, a tree frog is known whose development is very different from the typical frog's. Most frogs develop in a pond. A single fertilized egg develops into a small aquatic larva called the tadpole, which then develops into the adult frog by the process called metamorphosis. In this sequence, the frogs must have access to a pond to lay their eggs, for the tadpole has gills, just as a fish does, and only as metamorphosis proceeds do lungs develop that permit air breathing. In the tree frog, however, there is no tadpole stage. Instead, a small frog hatches from the developed egg. This is the same type of variation in life history that is found for the giant ants. No one knows how long it took for the alterations seen in the tree frog to occur or why they did.

Gigantism on the scale seen in *Them!* also seems to be a very unlikely outcome from the presence of radiation in the desert. The giant ants of the film have gone way beyond the increase in size that we call gigantism in humans. Human giants are on the order of 1.5 times taller than normal. The ants are more like 150 times longer than normal. This degree of gigantism poses problems that are not likely to be overcome without a whole collection of mutations. It is very unlikely that several mutations would occur at once to produce a viable giant ant.

Consider the problems of a giant ant. Its first big problem is that it probably cannot get enough oxygen to survive. Ants do not move air actively the way humans do. Instead, it diffuses passively through their airways. The diffusion time for oxygen changes with the square of the distance to be covered. That means that an ant about 1-inch long will have diffusion times 10,000 faster than an ant that is

100-inches long. The needs of each cell would be about the same, and it seems very unlikely that a 1-inch ant enjoys a safety factor as big as 10,000 in its supply of oxygen. As a result, we would predict that at best a giant ant would be very sluggish in its activities and at worst it would suffocate from lack of oxygen.

Another problem is the exoskeleton that provides support and protection for an ant. Exoskeletons are made of a polysaccharide material called chitin, and it is doubtful that chitin is strong enough to support bodies that are so much larger and heavier than normal. The other arthropods that use chitin in their exoskeletons are all relatively small creatures, including crabs, lobsters, and shrimp. The largest animals on Earth use endoskeletons for support, as we do. The extinct dinosaurs had endoskeletons, too.

Whales are mammals, as we are, and they have endoskeletons for support. But when a whale is beached and no longer has the buoyancy of water to help support its body, its skeleton cannot support the whale any longer. This illustration demonstrates the difficulty of making drastic transitions all at once. It seems very unlikely that the ants could suddenly accommodate their new massive bodies any more than the whale could tolerate leaving the water.

While we have explored how unlikely the giant ants in *Them!* are in real life, this does not mean that radiation is to be taken lightly. Survivors of the atomic bomb attack on Japan have high incidences of cancer and reproductive difficulties. People who were exposed to radiation in the Chernobyl accident are also experiencing radiation-induced illnesses. The ominous tone of the film is appropriate, for we know only of negative consequences from increased levels of radiation in our environment.

EXERCISES

1. Scientists sometimes use ultraviolet light or chemicals to produce mutations in the organisms they are studying. The mutated progeny have DNA that differs from their parents' DNA. There are mutations that occur spontaneously, but their occurrence is extremely rare. What evidence is presented in the film that the giant ants were mutants and that they originated from normal ants that had been exposed to radioactivity? The biggest nuclear reactor accident in history occurred at Chernobyl and exposed many humans to much more radiation than they would normally experience. Do you know of mutations in babies born after this accident? How long would it be before they would be seen? Do you know of other problems from the exposure to radiation in the Chernobyl accident?

2. In the case of the tree frog that has no tadpole stage, there is an advantage in skipping the tadpole stage. No longer must the tree frog live at a pond's edge. The range of the animal can then be extended and need not be limited by a need for a pond. This could be a valuable capability and allow the frog to move away from predators, for example. What other advantages might there be in being freed from the need to live at the pond? Can you think of advantages for the ants with their development altered? Can you think of advantages for the ants when their

size becomes so much larger? What are the disadvantages? On the whole, do you think the advantages outweigh the disadvantages or vice versa?

3. Which types of cellular reproduction occur in normal-sized ants? Do you think they are the same for the giant ants? Why or why not?

4. Among ants, males are haploid, and females are diploid. Devise scenarios for an ant colony in which a mutant allele for giant size has appeared. First, what do you envision if the allele is dominant? Second, what would happen if the allele is recessive? Do you think the colony would survive in either instance?

Chapter Twelve
The Energy Needs of Living Things

Work Requires Energy

Everyone is accustomed to the idea that energy must be expended in order to accomplish work. It is easy to recall the effort expended, and thus the energy used, to do many daily tasks. Whether one prepares a meal, runs for the bus, walks the dog, sorts laundry, vacuums the carpet, or goes shopping, energy is invested in the activity. One useful definition of **energy** is **the capacity to do work.** As multicellular organisms, we use our bodies' cells to accomplish our work. In the examples of household work, the most apparent work is accomplished by the actions of muscles being controlled by the nervous system. Their activities, however, require support from other systems to supply oxygen via the blood and lungs, to remove carbon dioxide and other wastes via the lungs and kidneys, to provide energy-rich food molecules via the digestive system, and to replace cells worn out by the wear and tear of working.

Unicellular organisms also must carry out work. An amoeba is a one-celled organism that lives at the bottom of ponds and seas. When it encounters a food particle, it must use energy to engulf the food by the process called endocytosis. While humans use energy to provide their bodies with a water supply, an amoeba in a pond has too much water around it. The water that continually enters the cell must be continually removed. An organelle called the contractile vacuole does this work for the amoeba. Therefore, as we consider both multicellular organisms and unicellular organisms, we recognize that living things have work to do. Consequently, they must have a supply of energy.

Organization Requires Energy

As we have seen in the preceding chapters, living things are exceedingly complex and highly organized. A cell is made up of a staggering number of different types of molecules that are associated in specific ways to form the organelles, which in turn are assembled into the cell itself. One of the lessons from physics that applies to this situation is found in the Second Law of Thermodynamics. According to the second law, the natural tendency in the Universe is toward greater disorder. All physical entities in the Universe, including molecules, spontaneously tend toward increasing disorder. This increase in disorder pertains to the Universe as a whole. Isolated parts of the Universe, however, can overcome this tendency by using energy. Cells, then, have a need for energy to provide for their ordered structures.

Living things, whether unicellular or multicellular, have a need for energy, a need that surpasses their ability to function without an extraterrestrial source of energy. One of our neighbors in the Universe is the source of a tremendous amount of energy that reaches Earth continuously. The Sun sends radiant energy to Earth, and the amount of that energy has been estimated at 10^{24} cal/yr. While this is a huge amount of energy, it is useful to living things only if they have a way to capture it.

At this point the insights from physics enter the picture again. The First Law of Thermodynamics tells us that the amount of energy in the Universe is constant, but that it can change form. This means that if the energy from the Sun can be provided to living things in a form they can use, the Sun can be the ultimate source of the energy living things need to maintain their organized structures and carry out their various kinds of work. Fortunately for all kinds of living things, the process of photosynthesis serves to trap the energy of the Sun in the chemical bonds of energy-rich food molecules which can be used by cells of all sorts in their internal sets of chemical reactions that provide energy for cellular activities.

Before we can study photosynthesis or other energy-related cell activities, we must consider the basics of how cells carry out their chemical reactions. The reactions that take place in cells can be expressed similarly to the way a chemist would describe reactions taking place in the laboratory in a flask. There are two categories of reactions: those that release energy as chemical bonds are broken and those that require energy to form new chemical bonds and build larger molecules from smaller ones. These are, respectively, degradative and synthetic reactions:

$$\text{degradation} \quad A-B \longrightarrow A+B+\text{energy},$$

$$\text{synthesis} \quad A+B+\text{energy} \longrightarrow A-B.$$

When chemists carry out a reaction, they can use very harsh conditions to make the reaction succeed. Chemists do things like heat chemicals in a gas flame or place the reacting chemicals in very strong acids. These are conditions that cells cannot tolerate. Cells need a way to carry out their chemical reactions under mild conditions and still have the reactions occur rapidly. The solution to this problem is to use **enzymes** operating in a sequential series of reactions called a **pathway.**

Enzymes and Pathways

An enzyme is a protein molecule which acts as a catalyst for a particular reaction. Catalysts speed up reactions, but they are not changed by them. At the end of a reaction, the catalyst is the same as it was at the beginning. Importantly for cells, enzymes catalyze reactions under mild conditions. Cells utilize enzymes in pathways to accomplish in a series of small steps what might be accomplished by a chemist in a single step under harsh conditions not tolerated by cells. This difference can be symbolized as follows. Say a chemist can accomplish the conversion of compound L to compound P:

$$L \longrightarrow P.$$

This same reaction might be accomplished in a cell under mild conditions by utilizing a pathway:

$$L \rightarrow M \rightarrow N \rightarrow O \rightarrow P.$$

Each step would be catalyzed by a different enzyme that is specific for that particular reaction. The illustrative pathway operates with four enzymes that convert compound L to compound P, the same overall reaction that the chemist might carry out all at once.

Adenosine Triphosphate

Another feature of the chemical reactions carried out in cells is that they use a chemical called ATP (adenosine triphosphate) to donate energy to reactions that need it. This is true also for activities that can more readily be seen as the work of the cell, such as endocytosis, active transport, protein synthesis, mitosis, and cytokinesis. ATP serves as an energy donor for the cell when a chemical bond holding the third phosphate group in place is broken. With simplified symbols the reaction can be understood easily:

$$A - P_1 - P_2 - P_3 \rightarrow A - P_1 - P_2 + P_3 + energy.$$

The breakdown of ATP to ADP (adenosine diphosphate) and phosphate releases energy. The beauty of this system is that ATP is found throughout the cell, and, therefore, can provide energy whenever and wherever it is needed. As long as a cell has a means of providing a supply of ATP, the energy needs of the organelles can be met.

Hydrogen and Electron Carrier Molecules

The last aspect of cellular chemistry we need to consider before examining pathways is the role of hydrogen and electron carrier molecules. Just as the name suggests, these molecules transport hydrogens (as atoms with a positive charge) and electrons around the cells. They are found in one of two forms, loaded with hydrogens and electrons or unloaded. The chemical names are cumbersome, but handy acronyms are easy to work with. The three hydrogen and electron carrier molecules we will deal with are NAD^+ (nicotinamide adenine dinucleotide), $NADP^+$ (nicotinamide adenine dinucleotide phosphate), and FAD^+ (flavin adenine dinucleotide). These symbols represent the unloaded or empty forms. When they are loaded with hydrogens and electrons, they are symbolized, respectively, NADH, NADPH, and $FADH_2$. In general, when molecules are being broken down, the hydrogen and electron carriers get loaded, and when synthesis occurs, they unload their hydrogens and electrons, and the reaction produces the unloaded form.

Photosynthesis

Photosynthesis is the process by which the Sun's energy is captured for the use of all living things, autotrophs and heterotrophs alike. The type of photosynthesis we will examine is carried out by all members of the kingdom Plantae and some members of the kingdom Protista. Some bacteria of the kingdom Monera carry out a different kind of photosynthesis. The reaction that summarizes photosynthesis is

$$6CO_2 + 6H_2O + \text{energy} \longrightarrow C_6H_{12}O_6 + 6O_2 \ .$$

In words, the reaction is carbon dioxide plus water plus energy yields glucose plus oxygen. Notice that the reaction combines two simple molecules that are readily available to form the energy-rich food molecule glucose plus oxygen. The radiant energy of the Sun is trapped and converted into the chemical energy present in the chemical bonds that hold the carbon atoms of glucose together in a molecule with six carbon atoms (a C-6 molecule) in contrast to the carbon dioxide molecules that have only one carbon atom (C-1 molecules).

Cells carry out photosynthesis in their chloroplasts with the process subdivided into two groups of reactions. In one set of reactions, called the **light reactions,** the Sun's energy is used to energize electrons which then move through the pathway to progressively lower energy levels. The energy released is captured in the form of ATP and NADPH. Water is split to provide positively charged hydrogen atoms and electrons which are needed in the pathway, and this results in the formation of oxygen molecules from the "leftovers." The molecules that carry out the light reactions are organized on the internal membranes of the chloroplast.

The second set of reactions is called the **dark reactions** because they do not use the sunlight. Instead they depend upon the products of the light reactions. The dark reactions are also called the **Calvin cycle** (in honor of one of the cycle's discoverers) as well as the **C-3 cycle** because a major part of the pathway uses C-3 molecules. The dark reactions begin with the joining of carbon dioxide (a C-1 molecule) to a C-5 molecule. This accomplishes the incorporation of one carbon atom into a larger molecule, so it is called **carbon fixation.** The C-6 molecule that is formed initially is unstable and immediately breaks down into two C-3 molecules. The cycle then carries out reactions that operate on the C-3 molecules and produces products useful for the eventual formation of glucose. When six molecules of carbon dioxide are fixed, enough carbon has been incorporated to account for one new molecule of glucose, a C-6 molecule. The cycle then regenerates the C-5 molecule for carbon fixation. In order for the cycle to run, a supply of ATP and NADPH is required—exactly the products of the light reactions. Thus we see that the light reactions produce molecules that are necessary for the dark reactions to run and that the dark reactions produce the energy-rich food molecule glucose. Its six carbon atoms are held together by chemical bonds which contain energy that was formerly radiant energy of the Sun.

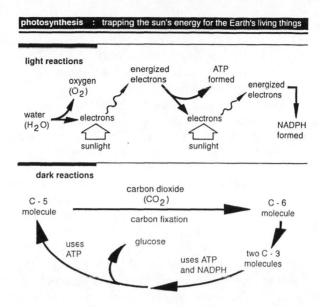

FIGURE 12-1. Photosynthesis. Two sets of reactions are necessary in chloroplasts to trap the Sun's energy in the energy-rich food molecule, glucose. The first set, the light reactions, uses water and the Sun's energy to form oxygen, NADPH, and ATP. The second set, the dark reactions, fix carbon dioxide into larger molecules which permit the cell to form glucose.

Glycolysis

The pathway that is the foundation of energy releasing pathways is called **glycolysis.** In this pathway, glucose, the energy-rich food molecule, has its chemical bonds broken which results in the release of the energy that was in the bonds. Some of the energy is captured in the form of ATP. Glycolysis breaks down the C-6 molecule glucose to two C-3 molecules of pyruvate. In addition to ATP, NADH is formed. When this pathway operates, nine enzyme-mediated steps convert glucose to pyruvate. Only a small amount of ATP is produced by the pathway—2 molecules for each glucose molecule split. Even so, cells, no matter which kingdom, and no matter whether they are the cells of a unicellular organism or a complex multicellular organism, depend upon glycolysis in the cellular scheme to provide ATP to meet energy needs.

The enzymes that carry out glycolysis are found in the cytoplasm. Note the fact that they have no need for oxygen. Because of this, glycolysis is an **anaerobic** pathway. Anaerobic processes need no oxygen; **aerobic** processes need oxygen.

Mitochondrial Pathways

Pyruvate may be broken down further in eukaryotic cells by the activities that take place in mitochondria. Initially, pyruvate is degraded to a compound called acetyl coenzyme A and carbon dioxide in order that the mitochondrion's **citric acid cycle** can degrade it. Also formed is NADH. By degrading acetyl coenzyme A, the citric acid cycle, also called the **Krebs cycle,** completes the breakdown of glucose to the final C-1 product, carbon dioxide. We humans think of ourselves as producing carbon dioxide as a waste product to be exhaled, and most of it is produced by the mitochondria completing the degradation of glucose. The Krebs cycle, like glycolysis, produces a little ATP as it operates (2 ATP molecules per glucose molecule). It also results in hydrogen and electron carriers being loaded up. The pathway yields two kinds of carrier molecules in "loaded" form, NADH and $FADH_2$.

The second pathway of mitochondria is the **electron transport system** whose operation produces the bulk of the ATP that results from the degradation of glucose. At this stage the loaded hydrogen and electron carrier molecules that were produced by glycolysis and the Krebs cycle are used to start the operation of the electron transport system. When it runs, ADP and phosphate are joined to form ATP. This is called **phosphorylation.** The combination of running the electron transport system and forming ATP molecules is called **electron transport phosphorylation.** It produces up to 34 ATP molecules per glucose molecule.

The operation of the electron transport system continues only as long as the molecule that accepts the electrons being transported is provided. That molecule is an oxygen molecule. As long as a eukaryotic cell has a supply of oxygen, the electron transport system can run by using the NADH molecules generated from glycolysis, the formation of acetyl coenzyme A, and the Krebs cycle, and by using the $FADH_2$ molecules generated by the Krebs cycle. If no oxygen is available, the system is blocked from further operation. Because the system requires oxygen it is **aerobic.** When we think of ourselves as needing to breathe oxygen into our bodies, it is primarily to operate electron transport phosphorylation and provide body cells with an adequate supply of energy. Eukaryotic cells need the large supply of ATP that is provided by this pathway because their complex activities demand much more energy than prokaryotic cells.

Silent Running

The film *Silent Running* underscores the seminal role of plants in the overall energy picture for living things on our planet. In the year 2008 when the story takes place, Earth is barren and can no longer support plant life. Giant space freighters have been rigged with huge domed greenhouses to maintain the plants of Earth. Each greenhouse has a collection of plants that occur together normally and grow under similar conditions. This project is a large-scale version of similar endeavors of botanical gardens, zoos, and universities in contemporary times. For reasons that remain unspecified, the crew is ordered to destroy the greenhouses and return to Earth.

Lowell, the crew member who has been responsible for the care of the plants in the set of greenhouses on one freighter, finds it impossible to destroy the plants.

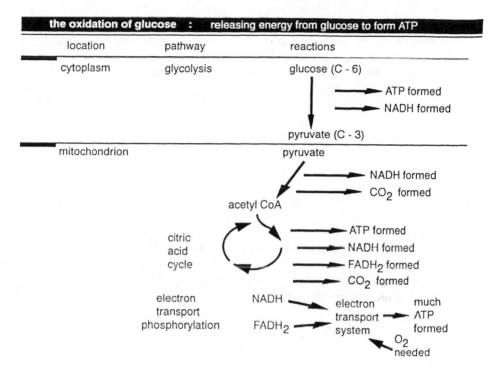

FIGURE 12-2. Releasing Energy from Glucose. In the cytoplasm glycolysis degrades glucose to pyruvate and forms ATP and NADH. Then mitochondria degrade pyruvate to carbon dioxide. Operating the electron transport system requires oxygen and yields large amounts of ATP

He disobeys orders, saves a greenhouse, and then attempts to mislead his superiors into believing he has crashed in the rings of Saturn. The bold mutiny appears to be for naught, because the plants begins to die. Lowell is despondent over the condition of the plants, and finally realizes, as a result of a comment by would-be rescuers, that he is in dim light. Saturn is much farther from the Sun than Earth, and the problem is magnified because light intensity varies with the square of distance. Since the average radius of Earth's orbit around the Sun is 93 million miles and for Saturn it is 887 million miles, the average intensity of sunlight at Saturn is about 1/100 the intensity of sunlight reaching Earth.

Plants must be able to tolerate variation in the amount of light reaching them because they must survive bright, clear days, light cloud cover (e.g., 100-m thick), and heavy cloud cover (e.g., 1-km thick). If the amount of light reaching plants on a clear day is 100%, then light cloud cover would reduce it about 50%, and heavy cloud cover would bring it down to about 10% of a clear day. Even so, that is still 10 times more than that reaching the plants in the domes in the vicinity of Saturn.

Moreover, the plants near Saturn get no relief from the dimness, while the clouds come and go on Earth. It is no wonder that the plants in the dome were in trouble.

Now let us examine why, from an energy standpoint, the plants were dying. The process of photosynthesis traps the energy of the Sun in the chemical bonds that hold the glucose molecule together. Excess glucose is used to form giant polymeric molecules of **starch** which stores the glucose until it is needed. When glucose is needed to provide the plant's cells with energy to carry out its activities and maintain its organized structure, starch is degraded to form glucose, and glycolysis begins the breakdown of glucose to form ATP and NADH. Then the pyruvate can be broken down further by the mitochondria of the plant's cells to form ATP by the Krebs cycle and by electron transport phosphorylation. Thus, if no photosynthesis takes place to form glucose, there is no supply of ATP to power the activities of the cell.

This lesson carries over to the welfare of the heterotrophic organisms, including ourselves. If there is inadequate trapping of energy by photosynthesis to produce energy-rich food molecules, then the heterotrophic organism dies because there is not enough food available to provide an adequate energy supply.

Plants supply heterotrophs like us not only with energy-rich food molecules, but also with oxygen. Geologists discovered that the appearance of oxygen in the atmosphere millions of years ago corresponded with the ability of photosynthetic organisms to produce oxygen during photosynthesis. We humans still depend upon plants to generate oxygen and replenish the atmosphere. About 80%–90% of the oxygen produced annually by photosynthesis is due to plants in the oceans. Terrestrial plants contribute about 10% of the oxygen annually. Thus, if Earth had become a place that was inhospitable to plants in 2008, it was also a place that could expect to support humans for only a limited time into the future. Let us hope that humans become good enough stewards of the Earth that no such time really comes.

EXERCISES

1. In the film *Silent Running*, would the plants in the domes have survived if the freighters had traveled no farther away from the Sun than the orbit of Mars? Support your answer with estimates of the percent of light from the Sun received by Mars in comparison to Earth.

2. In *Silent Running*, why might the officials on Earth have decided to abandon the plant specimens in space? Could life on Earth as we now know it survive without the kingdom Plantae? Could life on Earth as we now know it survive without any photosynthetic organisms? Explain your answer.

3. A local chapter of the Audubon Society sent out mailings recently which said on the envelopes, "Don't let the natural world become a memory!" Can you envision how humans could provide for their energy needs as the only surviving species on the entire planet?

4. Use your knowledge of photosynthesis to identify what changes on Earth might have threatened the survival of the plants by interfering with their ability to conduct photosynthesis. Which parts of the photosynthetic pathway would be affected by the changes you propose? Are any of the changes you suggest currently underway on Earth? What changes do you know of that could have an impact on photosynthesis? What is meant by the "greenhouse effect"? How would it affect photosynthesis?

Chapter Thirteen
Plants and Animals Compared

In this chapter we will consider some of the organizational characteristics in the two multicellular kingdoms that are most obvious and prominent to us: Animalia and Plantae. So far we have examined organisms mainly at the subcellular and cellular levels. In order to progress to higher, more complex levels of organization, we must examine the route that leads beyond existence as a unicellular organism to the multicellular organism.

Differentiation

By studying cell structure and function, we have seen the great similarity of all living things due to their cellular composition. However, the cells of multicellular organisms are able to conduct their various separate duties because of their differences; they become specialists with the ability to carry out certain tasks but not others. Even though the cells of most species all contain the same genetic information, they develop in ways that lead ultimately to the distribution of an organism's tasks to specialized cells. The cells are called differentiated cells, and the process that produces them is **differentiation.** Both your white blood cells and your skeletal muscle cells are human cells and contain the genetic information for humans, but these two cell types are very different in both structure and function. Multicellular organisms have their differentiated cells organized in ways that permit them to perform the variety of activities that are necessary to maintain the life of the organism. Now we will follow the levels of organization scheme through the tissue, organ, and organ system levels to see how this life strategy works.

Tissues of Animals

A **tissue** is a collection of cells that are similar in structure and function. In animals there are four kinds of tissue: **epithelial, connective, muscle,** and **nervous.** All epithelial tissues share the function of forming the surface of body structures. They all have a side that has no contacts with other cells. Further, because epithelia act as barriers and boundaries, the cells lie very close to each other and usually are held tightly next to one another by special junctions. Epithelial cells can be categorized further according to the shapes of the cells, e.g., flat, cuboidal, or columnar.

Connective tissues all have the function suggested by their name; one way or another, they provide connections to other tissues. They differ markedly from epithelia in that they usually secrete large amounts of extracellular material. There is typically a comparatively great distance between cells because of the extracellular material. Examples of connective tissues are tendons that connect muscles to bone, blood that connects all parts of the body with one another, and cartilage that caps the ends of bones to form a smooth surface for joint movement.

Muscle tissue is specialized to be able to contract. All muscle tissue, whether it is found in skeletal muscles, the heart, blood vessels, airways, the digestive tract, etc., has the ability to shorten, and thereby to apply a pulling force or to control the size of various structures. Blood pressure is controlled in part by the contraction of tiny bits of muscle tissue in the blood vessels. When the muscle cells contract, the diameter of the blood vessel decreases and the blood pressure increases. When the muscle cells relax, the diameter of the blood vessel is increased and blood pressure drops. The mixing of food that is being digested in the stomach and the small intestine is due to the contraction of muscles in the walls of those organs.

Nervous tissue is specialized for the transmission of impulses that convey information from place to place in the animal body. The function is one of communication and coordination. Cells called neurons carry the impulses. If you prick your finger on the thorn of a rose, the impulse is carried from your finger to your brain. There the impulse is transferred to the brain centers that permit you to realize what happened. Other impulses will signal your muscles to remove your finger from the thorn.

Organs of Animals

Two or more tissues associated to perform a common task and organized as an anatomically distinct structure form an **organ.** Organs are quite familiar as components of animal bodies. Everyone can probably call to mind several organs easily. When lunchtime nears, perhaps the organ that springs to mind will be the stomach, an organ specialized to process food by reducing it to a sloshy mixture, much the same as a quick run of small pieces of food and liquid through a blender achieves. This organ includes three types of tissue which are arranged to form three functional layers. The various layers are held together by connective tissue. The outer epithelium secretes a slippery material that permits the stomach to slide over neighboring structures as it moves. The middle layer is muscle tissue, mentioned earlier, which causes churning and mixing of the stomach's contents. The inner epithelium is specialized to secrete the digestive juices as well as mucus to coat the epithelium and thus to prevent the stomach from digesting itself. When the mucus protection fails, the stomach wall becomes injured. Its owner has a stomach ulcer.

Not all animal organs are quite so familiar. Bony fishes have an organ called a swim bladder whose function is to provide buoyancy. Spiders have an organ at the rear of their bodies that spins the silk of the web. It is called the spinneret. Within the animal kingdom we do find many familiar organs with broad distribution, but some of them are restricted to a single group of animals. Unusual organs normally confer on their owners a valuable ability that supports their way of living.

Organ Systems in Animals

A set of organs that carries out the same or closely related functions forms an organ system. The stomach is part of the digestive system. Its job is mainly to mix food with fluids. This is only part of the overall function of providing nutrients, water, and minerals to the body. The stomach works in cooperation with the mouth, lips, tongue, teeth, salivary glands, throat or pharynx, the esophagus, the small and large intestines, the liver, the gall bladder, and the pancreas. (See Table 13.1.) The human body also has a skeletal system, a muscular system, a nervous system, an endocrine system, a blood circulatory system, a lymphatic and immune system, a respiratory system, an excretory system, an integumentary system, and a reproductive system.

From Organs to Tissues in Plants

As we turn our attention to plants, we will consider first the organs of the dominant type of land plant, the flowering plant or angiosperm. A flowering plant has two basic body regions, the **shoot,** which is above ground, and the **root,** which lies below the ground. The shoot is composed of the stem and leaves. The **root,** the **stem,** and the **leaves** constitute the organs of the vegetative (as opposed to reproductive) flowering plant. Just as organs of animals contain more than one tissue, so do plant organs. For example, in a leaf there is epidermal tissue on the outside which protects the leaf. There is also ground tissue beneath in which the photosynthetic parenchymal cells are located, and vascular tissue embedded within the ground tissue. We recognize tissue groups as another level of organization in plants that lies between the tissue level and the organ level.

In flowering plants, whether they are towering cottonwoods or tiny violets, there are three **tissue systems.** (See Table 13.2.) **Dermal tissues** form the outer covering of the plant. They serve the plant much as skin does a human by providing protection and a barrier to water loss. **Vascular tissues** conduct fluids, primarily, but also add structural support to the plant body. **Ground tissues** lend strength and support and include cells responsible for photosynthesis and starch storage.

Plant tissues are more difficult to categorize than animal tissues, and as a result, there is some disagreement among botanists over how to classify plant cells by tissue. In a widely accepted approach, the tissue systems described above can be further subdivided into the actual tissue types of flowering plants. The major tissue of the dermal tissue system is **epidermis.** The epidermis protects the plant from injury and helps prevent water loss by forming a waxy cuticle on stems, leaves, and other exposed structures.

There are three major tissues in the ground tissue system; they are **parenchyma, collenchyma,** and **sclerenchyma.** Parenchyma is a category whose cells perform some diverse functions for plants. They include photosynthesis, starch storage, and regeneration to repair wounds. The green photosynthetic regions of

TABLE 13.1. Non-reproductive organ systems of humans.

Organ system	Organs	Functions
Endocrine	Glands: pituitary, thymus, thyroid, parathyroid, pancreas, adrenals, ovaries, testes	Release into blood various hormones
Blood circulatory	Heart, arteries, arterioles, veins, venules, capillaries, blood	Pump blood to lungs and throughout body to exchange gases, nutrients, & wastes
Respiratory	Nose, pharynx, larynx, trachea, bronchi, lungs	Exchange gases with air and blood
Lymphatic and immune	Lymph nodes & other lymphoid organs, thymus, bone marrow, white blood cells	Maintain fluid and osmotic balance, provide defense from foreign organisms
Digestive	Mouth, pharynx, esophagus, stomach intestines, salivary glands, pancreas, liver, gall bladder	Ingest and process food, eliminate waste
Excretory	Kidneys, bladder, ureters, urethra	Eliminate waste, maintain fluid and osmotic balance
Integumentary	Skin and associated organs	Protection, water retention, temperature control, sensory monitoring environment
Nervous	Brain, spinal cord, ganglia, nerves, organs of special sense	Receive, process, and send impulses to coordinate and integrate body functions
Muscular	Skeletal muscles	Movement and body support
Skeletal	Bones, ligaments	Support, motion, protection

TABLE 13.2. The tissue systems of flowering plants.

Tissue system	Major tissues	Functions
Dermal	Epidermal	Prevent water loss, protect from injury, control gas exchange, water storage
Ground	Parenchyma	Photosynthesis, storage of starch and other nutrients, repair of wounds, regeneration
	Collenchyma Sclerenchyma	Provide support to the plant body Protect plant structures from mechanical injury
Vascular	Xylem Phloem	Transport of water and minerals Transport of food

plants are familiar and contain parenchyma differentiated to perform photosynthesis. A potato has many parenchyma cells that are specialized to store starch. Collenchyma tissue has very thick, uneven cell walls. Cells of this tissue form fibers that lend support to the plant body. The "strings" of celery stalks are collenchyma. Mint plants have characteristic square stems, and it is collenchyma at the corners that gives them that shape. Sclerenchyma tissue is distinguished by its hardness. Sclerenchyma cells develop very thick, hard cell walls as they mature, and then many types die. An important component of the cell walls of many sclerenchyma is the compound **lignin.** Sclerenchyma cells form the pits of fruits, the shells of nuts, and the gritty stone cells of pears. They also form fibers that are sometimes useful to humans as well as plants. It is this type of schlerenchyma that forms the fibers of linen and ramie fabrics and also of the hemp that is used to make ropes.

The vascular tissue system is composed of **xylem** and **phloem.** Xylem transports water and minerals from the roots to other plant structures. The cells that form xylem synthesize very thick walls and then die to create the plant's "water pipelines." Phloem vessels are made of living cells. They transport sugars and proteins from regions where they are being formed to areas where they will be used or stored. For example, sugars are formed in the photosynthetic parts of the plant and transported via the phloem to the nonphotosynthetic parts that need the sugars for energy and other purposes.

In plants the three systems of tissues have their characteristic functions that prompt this way of categorizing them. In the vegetative plant they also have a characteristic anatomic pattern of occurrence. If one sections a plant serially, the tissues of the tissue systems are found to be continuous from leaf to stem to root. (See Fig. 13.1.)

Notice that in plants, **tissue systems** are a level of structural organization. The

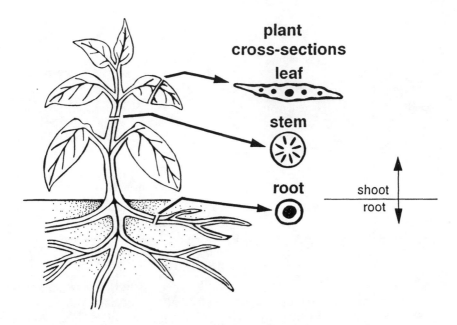

FIGURE 13-1. The major anatomic regions of a vegetative angiosperm and the characteristic locations of the three tissue systems. Black in an external region represents dermal tissue; in an internal region, vascular tissue; clear represents ground tissue.

system concept at this level of organization in a plant may be viewed in much the same way as it is in an animal. A group of organs in an animal is regarded as belonging to a system because of shared function. Neither anatomic proximity nor connectedness is required. Consider the endocrine system, which includes the endocrine glands of the human body. (See Table 13.1.) These glands lie widely separated in the body, but they share the role of forming and releasing in controlled fashion into the bloodstream the body's hormones. In plants the vascular tissue system includes different types of vessels that are anatomically separate, but they share the function of providing a liquid transport system to the plant.

Most often the organ system concept is not used for plants, and the tissue system is preferred. The continuity of the stems and roots and the repetitive structure of the leaves prompt this approach. However, the organ system concept could be applied. It would be similar to the integumentary system in animals, which is composed of the skin. The skin is a large continuous structure, which may be viewed either as the sole organ of the system or the main organ which incorporates smaller organs such as hair follicles. If the animal integumentary system

TABLE 13.3. Levels of Organization in Plants and Animals.

Level of Organization[+]	Animals	Plants
Organism	X	X
Organ System	X	(X)*
Organ	X	X
Tissue system		X
Tissue	X	X
Cell	X	X
Organelle	X	X
Molecule	X	X

[+]The levels of organization are arranged from the simplest components at the bottom to the most complex at the top. Each level is composed of structures from the level immediately below. For example, tissues are assemblages of cells that are similar in structure and function, and cells are assemblages of organelles that perform various functions. X indicates a level's presence.
*The organ system level is not usually employed in describing plants because it is neither especially, necessary nor helpful in studying plants. However, it is possible to construe plant anatomy in terms of organ systems.

can be a system with only one organ, then plant organ systems could also. We could speak of the root system or the leaf system. Other approaches are possible besides this one.

As we compare the levels of organization in plants and animals, we are in a position to see that the schemes are created to help us better study and understand the living thing. They are artificial devices, not a universal organizing principle found in the natural world. Thus, in plants the system level of the concept is most usefully applied after the tissue level, while in animals it is most useful after the organ level.

The Day of the Triffids

The film, *The Day of the Triffids*, tells the story of the aftermath of a mysterious meteor shower that somehow blinded people and caused a genetic alteration in certain plants, *Triffidus celestis*. Some triffids, as they were called, had come to Earth on meteorites many years before. Apparently radiation from the meteor shower mutated the triffids, which then grew into giant organisms that could move from place to place, killing humans with poisonous stingers on their tentaclelike limbs. They walked on their roots. Blind humans were no match for triffids, and the sighted humans did not do well against them either. To make matters worse, the triffids were reproducing. Like many other plants, they formed seeds that were dispersed on the winds. Just when it looked like triffids might conquer the Earth, two humans accidentally discovered that seawater killed the triffids. The dramatic discovery makes it clear that the triffids were not to win.

The triffids are supposedly plants, but in some ways their behavior is much more like animals. They therefore provoke a comparison of real, rather than fictional, plants and animals. First, it is not at all typical of plants to move from place to place. Plants have a very different life strategy than do animals. As autotrophic organisms their nutritional needs include small molecules and light. The small molecules can be obtained from the air around the plant or from the soil. Plants, therefore, use their roots to harvest nutrients from the soil and to secure the plant in place. Even though it is an interesting scenario to ponder, there do not seem to be plants on Earth that haul themselves free from the soil if their light is blocked or some other condition warrants seeking a new piece of real estate. Seed-producing plants generally form many seeds, and the ones that fall in desirable spots where needs are well met are the ones that thrive.

Animals, in contrast, have an approach to life that finds most kinds of them actively seeking their nutrients. This is especially evident among the land-dwelling (terrestrial) animals. They move from place to place in search of food, and they use muscles to move their bodies. Let us pay particular attention to animals with backbones—the vertebrates—which include fish, amphibians, reptiles, birds, and mammals. Use your own memories to envision vertebrates in search of food. You have probably seen squirrels gathering nuts, robins pulling worms from the ground, horses grazing in a field, and fish or frogs eating flies. In each case the animals achieve their movement by using skeletal muscles and bones in cooperation. The bones support body parts and provide places called joints for movement to occur. A skeletal muscle is an organ which contains muscle tissue that contracts as well as connective tissue to hold the various regions together and attach the muscle to bone. When a muscle contracts, it gets shorter and causes movement at a joint. This simple principle is used over and over again in animal movements, whether the movement of limbs for walking, the mouth for eating, or some other activity. Figure 13-2 shows how the shortening of a muscle in your arm allows you to bend your elbow.

Plants do move. Their movements are, in general, much slower than the movements of animals. When plants move, they do so by mechanisms that are totally different from those employed by animals. Remember, plants have no muscle tissue or muscles. One mechanism that is used by plants is unequal cellular elongation. When a plant grows in the direction of the light, the movement is accomplished slowly by forming longer cells on one side of the plant than the other. This process produces bending toward the shorter side. Another mechanism involves the movement of water and minerals to alter the rigidity of certain structures. This is the way that the leaflets of the sensitive plant, *Mimosa pudica*, close when they are disturbed.

Cellular elongation is the mechanism used by a venus flytrap to capture its prey. It accomplishes this task in a period of 1–3 s. This is a remarkable process, because the plant must elongate not only its cell membrane, but also its cell wall. Even though this is very fast for plant movement, it is still slow compared to the fast contractions that some muscles perform. Some muscles of the human body, such

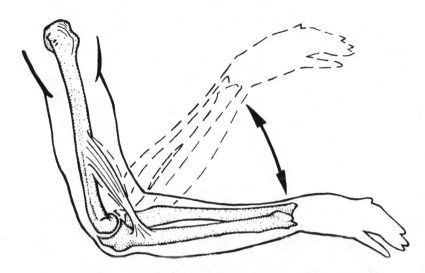

FIGURE 13-2. Bending the elbow is due to the contraction of muscles which pull on the forearm. The muscle shown to illustrate this concept is named brachialis.

as those that control the eyes, can contract in 0.001 s. Many other human muscles respond more slowly, but even the slower muscles respond on the order of 0.1 s rather than the whole second time frame of real plants.

Triffids, then, are much more animal-like in their speed as well as the manner of their movements. In moving from place to place they behave like animals. When plants move, they move some body part that is above ground; the entire organism does not move. The lightning-quick sting that kills the triffid's prey is another type of movement that is too fast for plants. It is animals that are capable of movements accomplished in less than a second, not plants.

EXERCISES

1. Some of the characteristics of triffids are plant-like and some are animal-like. Compile a table of the triffids' characteristics and determine whether they are plant or animal features. Use the traits mentioned already and try to identify others as well.

2. Imagine that a triffid has been captured alive. What observations would you like to make to characterize it better? How could you determine whether its organs are more plant-like or animal-like? What would you look for to determine whether its levels of organization are animal- or plant-like?

3. Next imagine that a triffid has been recovered shortly after its death. What studies might be performed to understand triffids more completely?

4. Is it plausible that seawater would kill triffids? Try to formulate an explanation for the lethal effects of the seawater on triffids.

5. What would you expect to find in a triffid if its locomotion is accomplished by animal-like mechanisms? What would you find if the movement is due to plant mechanisms?

Chapter Fourteen
Multicellularity and Immunity

Multicellularity

One of the major organizational distinctions among the kingdoms of organisms is whether the individuals of the member species exist as multicellular organisms. The three kingdoms in which multicellular organization is the norm are Fungi, Plantae, and Animalia. When the individual organisms have many kinds of differentiated cell types in the repertoire of the species, it is also necessary for the cells to have means of communicating with each other during the development of the individual and throughout the life of the organism. During development it is necessary for cells to receive and send cues that allow cells to group together properly or to segregate, and sometimes, as the body enlarges and takes shape, to migrate farther apart. The plants and animals are more likely to have such needs because they have more differentiated cell types and more orderly, fixed body plans than do fungi. The ways that animals carry out these signaling activities during their development are just beginning to be discovered. In comparison to animals, very little is known of such activities in plants.

Cell Adhesion Molecules

In animals we now recognize that the membranes of cells contain molecules that permit cells to adhere to one another or to the extracellular material. A molecular level recognition by membrane molecules of another molecule guides cells to associate with each other or to migrate over the extracellular material. The molecules that are involved in such cellular communication are called **cell adhesion molecules** or **CAMs.**

Still earlier in the development of animals we know that cellular adhesiveness is an important factor in embryonic development. Several kinds of animals have been studied to investigate the changing patterns of cellular adhesiveness. One of these is the sea urchin, a spiny marine creature that dwells in the intertidal zone. Sea urchins are like many other animals in that they start their development by forming a **blastula.** A blastula is a spherical embryonic stage formed when the one-celled embryo, the **zygote,** undergoes mitosis repeatedly. The blastula is not solid; it has a fluid-filled interior and is sometimes described as the hollow-ball stage of development.

The next step in the sea urchin's development must accomplish a dramatic change in the anatomic shape of the embryo. This next process is called **gastrulation,** and the embryo that results is called the **gastrula.** Its structure is more complex than the blastula and includes the presence of layers of similar cells to form the three **germ layers** of the embryo: **ectoderm, mesoderm,** and **endo-**

derm. This is the beginning of developmental processes in which cells sort themselves out in accordance with their future fate in the embryo. The fates of cells in each of the three germ layers are well known. The body covering will be formed from the ectoderm, the mesoderm forms skeletal structures, and the endoderm forms the gut. To accomplish the formation of the germ layers, some cells of the blastula must lose their affinity for their neighbors in the wall of the blastula and the layer of extracellular molecules, called the hyalin layer, that surrounds the embryo. They must at the same time acquire an affinity for the extracellular molecules toward the interior of the blastula. Then these cells migrate into the embryo, and gastrulation is underway. This is only one example of a situation in which the cell surfaces of developing multicellular animals must be able to communicate with other cells and the extracellular materials. It has been studied in great detail for a few animals, like the sea urchin, whose development can be conducted with some ease in a laboratory. Far fewer studies have been done on human gastrulation; nevertheless, humans go through the process of establishing germ layers, just as most animals do.

Histocompatibility

Among animals there is a form of cell-to-cell communication that permits the cells of an organism to recognize the other cells of that particular individual. This ability is so basic that it occurs even in the most primitive, simplest members of the animal kingdom—the sponges. If a piece of a sponge is broken off, it has the ability to regenerate. A simple regeneration experiment demonstrates the ability in sponges to recognize **self.** If two pieces of the same sponge are placed in contact as they regenerate, the two ends will grow together. However, if pieces from two different sponge individuals are placed in contact, they will not grow together. Only cells with similar self character will associate in the same individual. It is typical of multicellular animals to be able to distinguish between self and nonself.

The membrane molecules that give cells this ability are called **histocompatibility molecules.** Much more is known about the histocompatibility molecules of mammals, especially mice and humans, than of other kinds of animals. In humans the histocompatibility molecules that have been characterized are proteins that have some carbohydrate associated with them, making them **glycoproteins.** They are part of the cell membrane, and there are many kinds of histocompatibility molecules. Some of them have a major role in determining the self character of cells, so they are called **major histocompatibility molecules;** others have a minor role, and they are called **minor histocompatibility molecules.**

A good illustration of histocompatibility molecules in action is the transplantation of skin. If a surgeon transplants a piece of skin from one part of your body to another part, that skin graft will heal in its new location. Because the graft is self tissue, it is recognized as such and is accepted as self. This fact has been demonstrated by the use of skin transplantation for burn victims for many years. If, however, skin is transplanted from someone else's body, the transplant will be rejected as nonself. This phenomenon can be seen as the transplanted skin fails to become connected to the host's blood supply, but instead is attacked by the immune system and starts to die. Eventually it will look like a scab and be sloughed off. The exception to this scenario occurs for a person who has an identical twin. In this circumstance there are two people with identical histocompatibility molecules on their bodies' cells. For them a skin transplant between individuals would succeed.

Everyone is becoming increasingly familiar with surgical transplantation. We read in the news about a variety of organs being transplanted—heart, lungs, kidneys, liver, and bone marrow— and other types of transplantation are in experimental stages, such as pancreatic islets for diabetics. The big problem is not the surgery itself, but the lack of identical histocompatibility molecules in the donor's and the host's cells. Individuals who receive organ transplants must also receive medication to stop rejection responses that would normally attack and kill nonself cells because it is so rare to be able to provide a donor organ that completely matches the host in histocompatibility character.

Immunity

The immune system has the job of attacking foreign cells and molecules. To the immune system, all cells and molecules exist in one of two categories—self, which is not to be attacked, and nonself, which is to be attacked. In the preceding examples the introduction of nonself character was accomplished artificially by surgical transplantation. In normal living, multicellular organisms encounter nonself character regularly, and the immune system deals with it. Most often the foreign character comes in the form of bacteria or viruses, but fungi and protozoa are also capable of entering animal's bodies and interfering with their ability to function normally or to survive. Many attacks are unnoticed by the host because the immune system is successful in repelling the invader. When you get sick, your immune system is having trouble with an invader. The battle is a little more difficult with some types of foreign material than others. Foreign material that can be specifically recognized by the immune system is called an **antigen.** It is the task of the immune system of an individual to recognize, attack, and kill or inactivate antigens.

Some immunity is provided by mechanisms that offer general protection against many kinds of potential microscopic invaders. This is called **nonspecific immunity** or **innate immunity.** Examples are the barrier formed by the skin, the mucus of the respiratory system that traps tiny organisms, the acid of the stomach which kills cells, the flushing action of tears or diarrhea, and the moderate acidity (low pH) of the urethra that prevents the growth of harmful microscopic cells.

Some antigens evade the nonspecific immune mechanisms, and then another defensive system comes into play. This system is called **specific immunity** or **acquired immunity** because it selectively responds to one type of invader at a time and is only triggered by exposure to that invader. Specific immunity is launched by the white blood cells called **lymphocytes.** There are two types of lymphocytes: T-lymphocytes (also called T cells) and B-lymphocytes (also called B cells). T cells initiate the cellular or cell-mediated immune responses which are the more primitive type of immunity that is found comprehensively distributed throughout the animal kingdom. B cells initiate humoral or antibody-mediated immunity, which is probably more familiar than cell-mediated immunity but less widely distributed and less primitive.

Both B cells and T cells have receptor molecules in their cell membranes and use them to recognize the molecular features of cells, viruses, molecules, or molecular complexes they encounter. If the lymphocyte receptor recognizes a target such that a molecular lock-and-key fit occurs, the lymphocyte is activated. When a B cell is activated, it is stimulated to proliferate. It undergoes mitosis that produces some daughter cells that differentiate into cells called **plasma cells.** Plasma cells secrete **antibody** molecules, which are also called **immunoglobulins.** Other daughter cells become **memory cells,** which will be activated at a later time if the same antigen is encountered again. A typical target for antibodies to attack is an extracellular bacterium or virus.

When a T cell is activated, it, like a B cell, is stimulated to proliferate. Also like a B cell, some of the daughter cells will be memory cells. Other daughter cells differentiate into active T cells that help to fight the antigen. These are called **effector T cells.** Effector T cells carry out their functions when they are very close to the cells and molecules with which they interact with. For example, a cytotoxic T cell, also called a T_C cell, upon recognizing a target cell, releases a chemical that kills it. The active molecules released by T cells are called **lymphokines.** A typical target for a T_C cell is a virally infected cell because when a virus takes over a body cell and uses it for viral reproduction, some of the viral molecules end up in the cell membrane of the invaded cell. Such molecules are antigens that can be recognized by the T_C cell. Cellular immunity is specialized to protect against altered cells with some nonself character and carries out defensive actions close to the target cell. The lymphokines released operate at short range and are produced in small amounts. Effector T cells differ from plasma cells with respect to their secreted active products. Plasma cells' products, the antibodies, are secreted in large amounts, and they travel great distances in the body by way of body fluids or "humors."

Another way in which cell-mediated immunity differs from humoral immunity is that cell-mediated immunity involves not only recognition of antigen, but also of self character. Thus, when a T cell is activated, it must encounter self character, in the form of histocompatibility molecules, and foreign character. If only self character is encountered, the T cell is not activated. It has in this instance detected a perfectly normal situation—a body cell with its personal identification molecules as part of its cell membrane. However, a virally infected cell and many kinds of cancer cells contain in their membranes molecules that are not normally encountered and are, therefore, considered nonself. Such a cell activates the T cell.

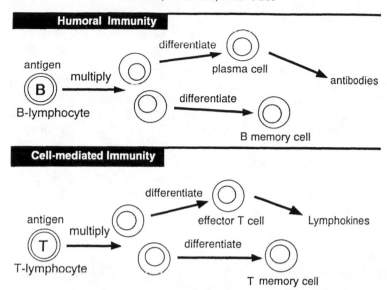

FIGURE 14-1. Specific Immune Responses. Humoral immunity is initiated by encounter with antigen. It leads to the production of memory B cells and plasma cells that release antibodies or immunoglobulins. Cell-mediated immunity is also initiated by encounter with antigen. It leads to the production of memory T cells and effector T cells that release lymphokines.

Fantastic Voyage

In this film a daring mission to save a scientist's life is undertaken by miniaturizing a submarine so that it can be injected into the scientist's bloodstream. Then the crew of the submarine plan to use a laser gun to destroy a blood clot in the scientist's brain. One challenge to the mission involves an attack on a crew member by antibodies. This is an interesting premise, but as we examine the character of humoral immune responses, you will see that it is very unlikely to have occurred. When an individual encounters an antigen for the first time, a **primary immune response** is elicited. If one measures the levels in the blood of antibodies that recognize the newly introduced antigen, they begin to appear in measurable amounts about three to four days after the antigen exposure. They then increase sharply until about seven to ten days after the exposure. Next the antibody levels start to decrease slowly for three to four weeks until they reach very low or even undetectable levels. If the individual encounters the same antigen again, a **secondary immune response** ensues. Antibody levels begin to increase very rapidly without a lag of a few days. The peak levels that are achieved far exceed those of a primary response. This very rapid response that is highly efficient at producing specific antibody occurs because of the participation of B memory cells.

The film shows antibodies attacking one of the tiny adventurers, but even a secondary response is not likely to have occurred so rapidly and produced adequate levels of antibody to protect against these invaders. It is certainly unlikely that the scientist encountered the "antigen" previously, but it is possible that he encountered something that had the same molecular shapes. Then this exposure would behave like a second encounter, even though it was really a first encounter.

The nature of antibody action is also misrepresented in the film. Antibodies themselves do not kill their targets, instead they earmark their targets for destruction by other participants in the immune response. One very important device for destruction is **phagocytosis,** the process in which a cell ingests particulate matter. Phagocytes are the cells that carry out phagocytosis. Among the phagocytes of the human body are the white blood cells called the neutrophils (also called polymorphonuclear leukocytes) and monocytes. Unlike neutrophils, the monocytes have not completed their maturation processes. They leave the blood and mature into cells called macrophages which are found throughout the body's tissues. When a target is coated with antibody, it is subject to rapid, efficient phagocytosis because the phagocytes carry membrane receptor molecules that recognize parts of the bound antibody molecules. Ingestion of a target by a phagocyte triggers processes that kill the target if it is alive, and then powerful enzymes degrade the ingested material.

A second important means of killing targets earmarked by antibodies is due to the action of **complement.** Complement is a family of 11 plasma proteins which remain inactive unless antigen-antibody complex activates the system. Once activated, a cascade of reactions occurs, just as a cascade of reactions leads to the formation of a blood clot. The complement cascade, upon completion, forms a complex of molecules that are inserted into the cell membrane of the target. Thus the integrity of the cell membrane is destroyed, and without a functional cell membrane, the target cell soon dies. The death can be very dramatic, with the target cell swelling up and bursting.

Even before the complement cascade reaches its end, during intermediate stages, activated complement protein molecules can aid in the elimination of the antigen because they promote **inflammation.** They do so because they act as chemical attractants for the blood phagocytes, especially the neutrophils. The activated complement proteins also increase the permeability of the blood vessels so that plasma moves out of the blood vessels. The plasma brings with it more antibody and complement.

EXERCISES

1. Make a list of the foreign material to which the scientist's body was exposed in the film; that is, how many potential antigens were introduced in a way that could have led to interaction with the body defenses. If something is concealed inside the submarine for the entire trip, it is not available for the immune system to detect. Do you think the humans would be antigenic in their diving gear?

2. Only Cora Peterson is attacked by the antibodies. Is that what you would expect? Explain your answer.

3. The only instance of phagocytosis comes near the end of the story. What are the circumstances involved? Make a list of other times the adventurers should have been in danger from phagocytic cells. Can you find a reason for their failure to be attacked?

Chapter Fifteen
Evolution

The theory of evolution is a central unifying concept in biology, just as the cell theory is. It is important to remember that in biology a set of ideas or concepts is called a theory only after a substantial amount of evidence has been gathered in its favor. At the time Charles Darwin published *The Origin of the Species* in 1859, he had gathered a tremendous amount of data to support his hypotheses, and in the years since, even more favorable data have been gathered by many scientists. Sometimes the theory of evolution is misunderstood and misstated, so it is important to understand it correctly. It is one of the most compelling scientific concepts ever enunciated. The eminent evolutionary biologist Theodosius Dobzhansky believed so strongly in the importance of the theory of evolution that he said, "Nothing in biology makes sense except in the light of evolution." In order to grasp and appreciate the Darwinian theory of evolution, some background to provide a context for it is necessary.

Environments and Prevailing Organisms Change

Once life arose on Earth, if it were to persist, it had to be able to survive under the prevailing conditions. That means that organisms had to be able to provide for their basic needs as individuals in order to live and reproduce their own kind. A success in reproducing a new generation then had to be reiterated again and again over extremely long periods of time, literally thousands, hundreds of thousands, and millions of years. Geologists have shown that, over such long periods of time, the climate at a given location on Earth has changed dramatically. So, too, have geologists shown, does the array of living things in a region.

Studying **fossils** has told us much about the living things from times long past. Fossils are the preserved remnants of the bodies of organisms that have been infused with minerals and whose anatomy (mostly external) is preserved as a rocky model of the original or the mineralized hard body parts themselves (such as a bone or shell). Another type of fossil is the preserved evidence of an organism, such as a footprint in mud. Fossils have revealed that over time periods that span hundreds of thousands and ultimately hundreds of millions of years, as the environment at a given spot on Earth changes, the ability of a given type of organism to survive is influenced. For example, the range of temperatures experienced in a year or the regional rainfall may change significantly. The fossil record tells us that the region that is now the Sahara desert, an arid and very hot place, was once home to lush tropical forests. But that was 15 million years ago and earlier. The dry climate that now prevails supports a completely different set of plants. The fossil record also shows that the region we now know as the plains of Nebraska was once covered by a sea. That period was 65 million years ago and earlier. Obviously a switch from an aquatic environment to a terrestrial one involves changes in the organisms occupying the region. Fossils found in both regions contain the remnants of former resident organisms. Fossils of ancient tropical plants that thrived in a moist environment have been found in the Sahara, while fossil marine organisms are plentiful in Nebraska's sedimentary rocks.

Recall that photosynthesis carried out by plants is responsible for the production of atmospheric oxygen. Prior to the accumulation of oxygen from photosynthesis, the conditions on Earth were predominantly anaerobic, meaning there was no supply of oxygen. Accordingly, the organisms that populated the Earth during those times were anaerobic. The fossil record and the chemical composition of rocks reveal that, with the appearance of cyanobacteria, life on Earth changed dramatically. Their type of photosynthesis was very good at producing ATP for their energy needs. Eventually the cyanobacteria, with their type of bacterial photosynthesis, outstripped the other photosynthetic bacteria with their older type of photosynthesis that did not produce oxygen. Earth would never be the same again. The anaerobic organisms would never again have much of the Earth to themselves. They became and remain inobstrusive organisms that live in anaerobic muds at the bottom of ponds, lakes, and seas or in a few isolated remnants of primitive environments such as the hot springs in Yellowstone Park. A change in the environmental conditions, then, can determine whether a species survives and whether it will be a major or a minor part of a region's ecology.

If life is to persist overall, living things must display a versatility and resilience to changing conditions. The fossil record demonstrates just such qualities in life, for it has existed on our planet for billions of years. Life has not merely persisted; it has gone beyond that and thrived. Living things have progressed from simple, single-celled organisms living in a world without oxygen to a complex array of organisms that live in all sorts of environments—aerobic or anaerobic, hot or cold, watery or dry, etc. Table 15.1 lists the major geologic time periods, their prevailing forms of plant and animal life, the dominant climatic character, and the major geological activity. They were not worked out in this amount of detail in Darwin's time; in fact, the scheme of geological continent building was not worked out at all.

Natural Selection by Survival of the Fittest

Charles Darwin spent years contemplating the succession of organisms shown in the geologic record, and he, himself, observed the distinctive sets of organisms living isolated on islands as he traveled around the world on a 5-year exploratory voyage. Slowly, he began to make sense of the patterns he saw. One thing he noticed was that organisms are very well matched to the prevailing conditions of the places where they live. It occurred to him that only the organisms that could deal with the conditions of a particular place survived and reproduced their kind; those that could not deal with the conditions died. He concluded that there is a continuing process of **natural selection** that takes place. The types of organisms that survive are the ones best suited to the time and place; they are the ones that

TABLE 15.1. Prevailing types of organisms, as revealed by the fossil record, and dominant climatic and geologic features during the geologic time periods.

Period	Time*	Features
Quaternary	2	Biology: humans evolve Climate: periods of glaciation Geology: North & South America join
Tertiary	65	Biology: Mammals diversify, flowering plants and gymnosperms (conifers) plentiful, grasslands and forests expand Climate: periods of glaciation, cooling in latter half of period Geology: continental separation continues, mountain building
Cretaceous	135	Biology: flowering plants arise and become dominant, large reptiles (dinosaurs) become extinct, ancient birds become extinct Climate: cools at the end of period Geology: mountain building, Africa & South America split
Jurassic	190	Biology: dinosaurs dominant animals, crustaceans common, fish diversify Climate: very little variety by latitude or time of year Geology: extensive inland seas
Triassic	230	Biology: Ferns and conifers dominant, dinosaurs and other reptiles first appear Climate: deserts widespread Geology: continent formation occurring
Permian	280	Biology: reptiles expand, many amphibians and invertebrates die out, conifers appear Climate: cool climate begins to warm Geology: mountain building, glaciation and dry climates
Carboniferous	360	Biology: amphibians dominant, reptiles appear, forests of seedless trees ("coal forests") Climate: warm and humid Geology: mountain building, glaciation

Continued on p. 204

*Times are in millions of years prior to the present and represent approximate times for the beginning of a period.

Period	Time*	Features
Devonian	420	Biology: land plants and first forests appear; insects, sharks, and amphibians appear; fish dominant Climate: periodic dry periods, cooler climates Geology: mountain building, fresh water basins form
Silurian ·	435	Biology: first land plants and animals, algae common Climate: vast swamplands Geology: extensive inland seas, slow uplifting of land
Ordovician	505	Biology: invertebrates diversify, early fishes, abundant algae Climate: mild Geology: shallow seas reach their maximum
Cambrian	570	Biology: first jawless fish, trilobites common, algae dominant plants Climate: warm climate Geology: extensive shallow seas near equator
Precambrian	2,500	Biology: origin of the eukaryotic kingdoms Climate: changed from hot, anaerobic conditions to cooler, aerobic ones Geology: major land masses present with shallow seas

*Times are in millions of years prior to the present and represent approximate times for the beginning of a period.

reproduce their own species. Darwin called this the **survival of the fittest.** He was writing about species, not individuals. In addition, his concept of fitness meant the ability to reproduce the species, not general physical prowess. Survival of the fittest, in Darwinian evolution, means that those species that are best suited to the local conditions will be the ones that survive long enough to produce the next generation of their species. The species that are successful are seen as being well adapted to the local conditions. **Adaptation** is another aspect of Darwin's theory.

Adaptation and Speciation

Adaptation occurring over long periods of time produces variation that can eventually form a new species. Darwin recognized this by observing the variation that could be obtained by controlled breeding of domesticated plants and animals. In the US, for example, we have some cattle, such as Holsteins, that have been bred to be excellent dairy cattle, while others, such as Black Angus, have been bred to be excellent beef cattle. They are not different species, but they are quite

beak structures of Darwin's finches

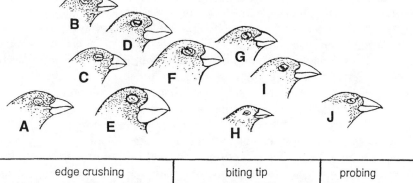

edge crushing		biting tip	probing
probing bills	crushing bills	grasping bills	probing bill
mainly plant food		mainly animal food	100% animal food

FIGURE 15-1. Beak structures in the finches Charles Darwin found in the Galapagos Islands (after T. Dobzhansky et al., *Evolution*. W. H. Freeman and Co., 1977). The various beak structures suit the birds to utilize different food sources. Each letter stands for a different species of finch.

different in some characteristics. The desired characteristics have been achieved by identifying animals which best display that characteristic and mating them. Their offspring are genetically enhanced for the selected trait. Repetition of breeding for a certain characteristic through many generations produces the distinctive strains desired by the breeders. Should they reach the point that they could no longer interbreed, then they would be considered different species. Types of organisms that cannot successfully interbreed or normally do not because of a factor like geographic separation form separate species.

Adaptations in the wild permit organisms to survive the rigors of climate, find food, escape predators, occupy or utilize an area that others cannot, etc. Darwin recognized that the finches of the Galapagos Islands were adapted to utilize different food sources. He identified 14 different kinds of finches in the islands, and they were different from the finches found on the mainland of South America. Figure 15-1 illustrates the specialized beaks found among some of Darwin's finches. Gradually Darwin realized that a single ancestral type could generate diverse descendants as the variation in a starting population became more and more specialized. The founding population of finches that came to the islands, like all populations, must have had variations in their characteristics, including beak shape. The birds with small, slender beaks were better at slipping their beaks into crevices and small holes in trees to find insects. The birds with larger heavier beaks were able to crush seeds and eat them. The birds feeding on insects in the trees would tend to breed with one another and maintain and enhance their advantageous beak structure. The birds with crushing bills that were eating cactus seeds would also tend to breed with one another. Eventually, the birds would never interbreed with the other type and each type would be specialized to utilize an available resource, rather than competing for the same resource.

The process of forming reproductively isolated populations is called **speciation.** A species, by definition, is a group of similar organisms that normally interbreeds among its members and not with other types of organisms. The offspring, in turn, must themselves be fertile. Not all matings that produce offspring produce fertile offspring. A familiar example of sterile offspring occurs when horses and donkeys mate. The mule is produced by breeding a male donkey to a female horse. Sterility occurs due to a variety of reasons. In this instance, the problem is a mismatch between chromosomes. The male donkey has 31 chromosomes in a sperm cell, and the female horse has 32 chromosomes in an egg cell. The mule ends up with 63 chromosomes in body cells. Because of the odd number of chromosomes, normal meiosis does not occur. The chromosomes cannot pair up and separate to form gametes.

The preceding definition of a species applies well to organisms that reproduce exclusively or chiefly sexually. It is not useful for recognizing species of organisms that reproduce asexually. This is especially true for micro-organisms that reproduce asexually. Bacteria, many protists, and some fungi are characterized as species instead by biochemical and physiological characteristics.

Homologies, Development, and Vestigial Structures

Besides the evidence for evolution found in the geologic record, evidence also occurs in the anatomic comparison of organisms both at mature stages and during embryonic development. The vertebrate animals (animals with backbones) are quite a familiar group and include fish, amphibians, reptiles, birds, and mammals. Observation of their embryonic and adult anatomy leads to the conclusion that they arose from a shared ancestor or ancestors and that evolution taking place over long periods of time produced the array of vertebrates now living on Earth. Skeletal anatomy is advantageous for study because of the great durability of bones. They are the body parts most likely to be preserved as fossils. As a result, the skeletons of present day vertebrates and also of ancient vertebrates, many now extinct, have been compared extensively. A basic vertebrate skeleton is seen in all groups, and its bones can be divided into two functional sets: bones of the axial skeleton and bones of the appendicular skeleton. The axial skeleton includes the skull, the vertebral column, and the thoracic cage (rib cage). The appendicular skeleton includes the bones of the limbs plus the bones of the pelvic (hip) and pectoral (shoulder) girdles that serve to attach the limbs to the axial body. It is remarkable, indeed, that all of the diversity in form and use of habitat among vertebrates occurs with such strong skeletal similarity. This similarity is known to

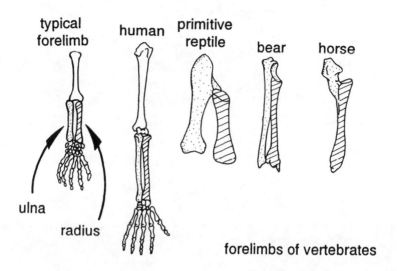

FIGURE 15-2. The forelimb of vertebrates, other than fish. The typical forelimb skeleton. The human forelimb in which the radius is the bone on the same side of the forearm as the thumb. The radius and ulna of a primitive reptile (after A. S. Romer, *The Vertebrate Body*, W. B. Saunders Co., 1962). The radius and ulna of a bear (after Romer). The radius and ulna of a horse (after Romer). The radius is drawn with diagonal lines on it; the ulna, with stippling.

extend to soft body parts as well for modern vertebrates, but such comparison is not possible for ancient types.

The comparison of skeletons demonstrates the variations in the basic anatomic plan that give distinctive character to the different types of vertebrates. One example is the bones of what, in humans, is the forearm. The general pattern among vertebrates other than fish is for this limb region to be formed from two bones, the radius and the ulna. However, their shape varies considerably, and they may even lose their separate identities and become fused to each other. Figure 15-2 shows the basic skeletal features of a forelimb in vertebrates other than fish. The radius and ulna are located between the upper limb region that has a single bone, the humerus, and the "wrist" made up of several small bones called carpals. In the human forearm the radius and the ulna are long slender bones; in a primitive reptile they are short and stout; in a bird's wing they are long and slender and have quite a lot of space between them. In a bear the two bones lie very close to each other, and in a horse, the two bones have fused to form one bone. Each animal uses its forelimb differently, and the differences are reflected in the anatomic conformations.

Because these various forelimb bones appear to be derived from the same original type of skeleton, the variations represent **divergent evolution.** The fossil

pectoral fin bones of fossil fish

Eusthenopteran type

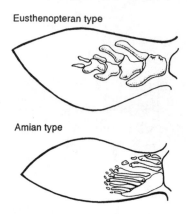

Amian type

FIGURE 15-3. Pectoral fin bones of fossil fish. A. the Eusthenopteran type (after Romer). B. the Amian type (after Romer).

record indicates that fish are the oldest vertebrate group, so one looks for the ancestral type among fish. Many types of fish have a large number of small bones involved in the skeletons of their fins. However, some do not; therefore, among these one looks for a skeletal configuration conducive to modification that could have produced the forelimb anatomy. Figure 15-3 depicts the left pectoral fin of a Eusthenopteran fish. Its arrangement of bones can be envisioned as giving rise to the forelimb bones with only slight modifications. This contrasts with other fin bone arrangements, such as the Amian type, that had multiple parallel elements.

The embryonic development of vertebrates also speaks in favor of a progressive alteration of a basic vertebrate body plan to yield, eventually, the modern groups of vertebrates we recognize today. For example, fish have associated with their gills a set of blood vessels called the aortic arches. Amphibians also have these blood vessels in their aquatic, larval stages. The blood in them exchanges oxygen and carbon dioxide with water flowing through the gills. However, reptiles, birds, and mammals have no functional gills; nevertheless, they form aortic arches during their embryonic development. The embryonic arches never function in gas exchange, they do not persist, and further development removes them.

Some organisms have organs for which no function is known, and they resemble in structure and location functional organs in other organisms. Two very interesting examples of such organs, which are called **vestigial organs,** are found in snakes and whales. Both animals have tiny thigh bones which appear to be without function. Neither animal has legs at all, yet each has very small thigh bones located in the position on the body where legs would be expected. These vestigial organs lead biologists to conclude that snakes and whales both had ancestors with legs. Divergent evolution produced some animals that continued to have legs as well as animals that lost their legs, but gained valuable traits that permitted their survival.

Planet of the Apes

The film, *Planet of the Apes,* was one of the first to make extensive use of newly developed realistic makeup techniques so that humans could be cast in the parts of apes of the future. What a topsy-turvy world is depicted, at least from the human perspective. No longer are humans the dominant organisms; instead, it is the evolutionarily advanced descendants of present-day chimpanzees and other apes occupying the position that humans do in our contemporary world.

Is it totally far-fetched biologically to envision a distant future, e.g., the year 3978 when the film's story takes place, in which apes have taken over the place of humans? To examine this question, let us compare chimpanzees and humans and also examine some aspects of human evolution.

One method of comparing organisms is to compare their DNA. Organisms that belong to the same species have highly similar DNA. All members of the species must have a functional set of genes with information to direct their development and to conduct the functions of their bodies' cells. If genes are highly similar, as is true within a species, the DNA, which is normally present as double-stranded molecules, can be fragmented, separated into single strands, and then allowed to reform double strands spontaneously. The process is somewhat imperfect, but even so, for a given species, the DNA reforms into double strands very successfully. If one compares the amount of double-stranded formation when DNA of two species is mixed, one can ascertain how similar their DNA is. If the DNA is very similar, it will be very successful in forming double strands, even though it comes from different species. If it is very different, double strands will form poorly.

When human and chimpanzee DNA are compared, they are found to be amazingly similar. The estimates vary somewhat, but are always in the high nineties, 96% to 99%. This amount of similarity of the genetic material, itself, suggests that it would not take too much alteration to produce human characteristics in chimpanzees. It also demonstrates that the organization of the genes and the regulation of their expression is critical to the outcome of embryological development. Theoretically, there is no reason why chimps and humans could not have identical genes for the housekeeping functions of cells and for structural molecules, but differ in the developmental genes that control anatomic development.

In the classification of animals, humans and chimpanzees are both members of the class Mammalia, the order Primates, the suborder Anthropoidea, and the superfamily Hominoidea. They belong to different families; humans, to the family Hominidae, and chimps, to the family Pongidae. Many of the evolutionary changes that were important for humans were changes that took place for the entire primate order and not for humans only. These include changes in the skeleton so that forelimbs functioned less and less in weight bearing, a shift from the sense of smell to the sense of vision as the primary source of information about the environment and conditions of the moment, development of a larger brain, and changes in behavior.

The development of a larger brain is especially important because it plays a role in the other changes. For example, without an enhanced ability to process and interpret visual information within the brain, the sense of vision could not have gained more importance among primates. If the brain had not changed to accommodate the altered use of body parts, the modified anatomies could not have made as much difference in the lifestyles of the involved species. Dr. Zira, herself a chimpanzee, explains in the film her theory about humans: that their brains are the source of their inability to speak and their limited manual dexterity.

Contemporary chimpanzees have a brain volume in the range of 280–400 cm^3, whereas for humans it is about 1,300–1,400 cm^3. Since the size of neural cells is similar for the two species, humans clearly have a much greater capacity for receiving and processing information and formulating and directing responses. Within the last 20 years, it was common to encounter statements asserting that much of the human brain is unused. However, the functions of areas previously thought devoid of functions have been discovered. Present thinking about the brain expects that all areas have functions; some remain to be discovered or clarified. Therefore, we cannot safely assume that humans simply have more unused brain volume than chimps. Humans must have more neural capacity through the intricate combinations of cells and their "wiring" patterns that determine the capabilities as well as the limitations of the species.

It may strike us initially as practically impossible for such changes to have occurred in the genetic information of both species without involving a myriad of alterations in genes. Were that the case, human evolution would be much more difficult to conceptualize. However, we can see a single developmental change occurring in the enlargement of the brain, one that would require very little alteration of genes. This change is called **neoteny**. Neoteny is, strictly speaking, the return to a lesser stage of development after a more advanced stage has been achieved. A classic example is the amphibian *Ambystoma* which reforms the external gills it had during larval aquatic development at times when adults find conditions on the land untenable.

As it is applied to evolutionary changes, neoteny means that the development of a species tends to retain some juvenile or embryonic characteristics. In the case of human evolution, we can regard the development of a larger brain case (the cranium) and brain as a neotenic change. Juvenile humans have proportionately larger heads than do adults. Thus one way to enlarge the brain is to slow down development so that what was once a juvenile state becomes the end point of development, the adult state. If the genes that program human development changed so that the human head grew faster than the rest of the body, then the head would end up proportionately larger. Or if the genes were altered so that the other body parts grew more slowly, the same effect would be achieved.

Humans are the most extreme of the primates in the percentage of their lives spent in preadult stages as opposed to reproductive adult years. For humans about 18 years are preadult and about 32 years are reproductive adult years; that amounts to 36% preadult years. For chimpanzees about 12 years are preadult, and 28 years are reproductive adult years; that is 30% preadult years. For lemurs, it is about 4 years spent as a preadult, and 14 years as a reproductive adult; the lemurs spend only 22% of their lives as preadults. If we were to include the time of development in the uterus, the figures would be even more striking. These figures suggest that evolutionary neoteny has been a factor in the emerging differences seen among the primates.

EXERCISES

1. Practice thinking about how natural selection operates by describing how modern giraffes, which have long necks and feed on the leaves of trees, could have evolved from a population of ancestors with shorter necks. Try next to formulate a scenario in which chimpanzees could have evolved as depicted in *Planet of the Apes.* Use the information in this chapter, preceding chapters, and additional sources, if necessary.

2. A mass extinction of organisms, including the dinosaurs, occurred about 65 million years ago. One theory for how this extinction came about proposes that a giant asteroid crashed into the Earth and its impact raised a dust cloud that obliterated so much sunlight that plant growth was drastically diminished and resulted in the extinctions. Compare this kind of sudden alteration in the environment with that suggested in the film. Is it plausible for the catastrophe to have advanced the chimps and done the opposite to humans?

3. What human characteristics do you think are most important in supporting the human dominance of Earth? Defend your choices.

4. *Planet of the Apes* takes place in the year 3978 or roughly 2,000 years from the present. Is that amount of time long enough for humans and chimps to have changed as depicted in the film? The first hominoids appear in the fossil record about 20–23 million years ago. About 6–10 million years ago the first hominids emerged. The earliest hominids belonged to the genus *Australopithecus,* which is now extinct. Modern humans arose around 300,000 years ago. In formulating your response, take into account also the experience of artificial selection by humans. The breeding of cattle was mentioned earlier. Other examples include the various breeds of domestic dogs and cats. Include an explanation of your rationale in your answer.

5. Chimpanzee's fingers are longer in proportion to their thumbs than are humans'. Is this difference likely to hinder chimps in attempting activities that humans perform easily? Make a list of 10 activities that require the use of your hands and evaluate whether they require precision or power. Do the same thing for chimpanzees.

6. Dr. Zaius says that he considers Taylor a mutant and expresses his concern that since one mutant has appeared, there are bound to be many others. Without

considering whether or not he actually believes his statement, evaluate the likelihood of such a situation. Does evolution proceed with the generation of similar
mutations? Could Taylor be, as he is called at one point, a "missing link?"

7. Explain what happened 2,000 years earlier in the Forbidden Zone. Why was
the ancient taboo established?

8. What advanced capabilities have the apes attained in their culture? Which
ones are missing?

9. Is it reasonable and probable that the events that changed humans and apes
would have been regressive only for the humans and progressive for all of the
various types of ape? Defend your answer.

10. Researchers trying to discover the similarities and limitations of chimpanzees, pygmy chimpanzees (bonobos), and other apes and monkeys find that their
capabilities have been underestimated heretofore. In the book **Shadows of Forgotten Ancestors,** Carl Sagan and Ann Druyan recount the results of research that
demonstrates among contemporary primates "friendship, altruism, love, fidelity,
courage, intelligence, invention, curiosity, forethought, and a host of other characteristics" that are usually attributed especially, sometimes solely, to humans.
They suggest that, "Perhaps our uniqueness is only this: an enhancement of well-
established, preexisting talents for invention, forethought, language and general
intelligence—enough to cross a threshold in our capacity to understand and
change the world." What interpretations of the evolutionary events in *Planet of the
Apes* become more probable if these findings are correct?

11. In humans the brain region that is comparatively much larger than that of an
ape is the cerebrum. Using neoteny as a mechanism, explain what would happen
to a chimp's brain during development to make it more like a human's.

FILM DESCRIPTIONS

Films With Literary Commentary

2001: A SPACE ODYSSEY

MGM (US), 1968, color, 139 min.

Credits: Producer and director, Stanley Kubrick; screenplay, Stanley Kubrick and Arthur C. Clarke; loosely based on "The Sentinel," by Arthur C. Clarke; cinematographers, Geoffrey Unsworth and John Alcott; special effects, Douglas Trumbull.

Cast: Keir Dullea (Commander David Bowman), Gary Lockwood (Frank Poole), William Sylvester (Dr. Heywood Floyd), and Douglas Rain (voice of HAL).

PLOT SUMMARY

2001: A Space Odyssey portrays the development of humanity from animal to "star-child" through the influence of an alien artifact. The film develops in three sections linked by the image of the black monolith. Presumably, the monolith is an artifact of an earlier and more advanced extraterrestrial civilization.

The first section, "The Dawn of Man," opens with a desert scene of eroded rocks and fleshless bones. A series of scenes then depicts a band of apes feeding, living in caves, drinking from a waterhole, and being threatened by another band. Suddenly a black monolith appears, wordless choral music swells, the apes stroke the monolith, and they suddenly discover the bone as a tool (a club). When next threatened by the other band, one of them uses the bone club to kill the most aggressive attacker. After this success, the ape-being throws the bone into the air, where it fades into images of a space satellite, then a space station, then a space shuttle in transit, emphasized by the image of a pen floating from the hand of a sleeping passenger. This sequence suggests the evolution of humanity from animal to tool-using near human to space-faring intelligence because of the intervention of the monolith.

Still in "The Dawn of Man," the space shuttle passenger, Dr. Heywood Floyd, enters the space station, where images of open, barren vistas populated by solitary humans or small, isolated groups symbolize the isolation of humanity and the barrenness of its culture even in 1999. Dr Floyd's meeting with some Russian scientists who ask him about the mystery of Clavius on the Moon leads into his visit to the Moon, where another black monolith has been discovered. This one supposedly has been deliberately buried, and dates back 4,000,000 years. It produces a very strong magnetic field. Several men take a Moon bus to visit the excavation site, where one of the men strokes the excavated monolith in imitation of the ape-men at the opening of the film, and a sudden electronic signal is heard, directed toward Jupiter.

The second section of the film, "Jupiter Mission 18 Months Later," reveals humanity still dependent upon and victimized by its tools. It focuses on Mission

Commander David Bowman, Astronaut Frank Poole, and the HAL 9000 computer aboard *Discovery* on their way to Jupiter to investigate the possible receiver of the monolith's signal. Also on the ship, but in hibernation to reduce the amount of food and oxygen needed on the journey, are the three men of the survey crew, to be roused when the ship nears Jupiter.

The onboard HAL 9000 computer actually controls most of the ship's functions; therefore, when HAL later malfunctions, it cuts off communication with Mission Control on Earth, kills the hibernating survey crew by deactivating their life support systems, and lures Frank Poole outside the ship where it uses the EVA pod to cut his oxygen line, killing him. When Commander Bowman takes an EVA pod out to retrieve Poole's tumbling body, HAL attempts to keep him from reentering the *Discovery*. However, Bowman, now the only survivor, blows the door off his EVA pod to enter the emergency hatch and shut down HAL, and thus saves himself and the mission.

HAL supposedly mimics most activities of the human brain, is the "most reliable computer ever made," "enjoys working with people," is "constantly occupied," and "acts like he has genuine emotions." HAL's malfunction and the murderous results remain unexplained, although as Bowman deactivates him, HAL admits that he has made "poor decisions recently." However, he claims to "feel much better now," until, as Bowman continues to flick switches, HAL says, "My mind is going; I can feel it." The sections ends with HAL shut down and a prerecorded briefing playing now that the ship is in "Jupiter space and your entire crew revived." The briefing reports the evidence of intelligent extraterrestrial life: it is the 4,000,000-year old black monolith found buried 40 feet beneath the Moon's surface near the crater Tycho, as depicted earlier in the film.

The third section of the film, "Jupiter and Beyond the Infinite," reveals humanity's evolution beyond the industrial (or space) age. It relies upon shifting, juxtaposed and sequential images of Jupiter, its moons, the starfield, *Discovery*, a huge black monolith floating in space near Jupiter, and strange bands of light, which finally collapse into the image of an empty room furnished in Victorian style. An EVA pod from the ship appears in the room, then Bowman in his spacesuit. Bowman ages, and the camera transmits several rooms, all empty. After a couple more images of Bowman aging, he appears in bed, ancient, bald, and gaunt, breathing harshly. Suddenly he raises a hand, points toward a black monolith at the foot of the bed, and his breathing ceases. The film continues with the image of a fetus, obviously human, but with larger eyes and head, floating in space. The images shift from the fetus to the monolith to a view of Earth from space to a starfield, and finally to the fetus among the stars, gazing open eyed toward the globe of Earth. As the first monolith assisted ape-beings to evolve into space-going humanity, the monolith on the Moon assisted space-going humanity to travel into the farther reaches of the solar system, and the final monolith completes the evolution of humanity into the star-child, ready to take its place in the wider universe.

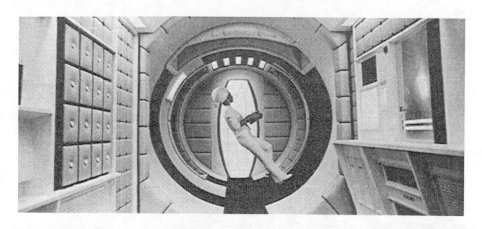

2001: A Space Odyssey. Stills 1–3 depict a stewardess walking upside-down in a shuttle. (Photos: Museum of Modern Art/Film Stills Archive. Courtesy of Metro-Goldwyn-Mayer.)

SPECIAL EFFECTS

2001: A Space Odyssey initiated the era of spectacular special effects in science fiction films. Actual photos of Jupiter and its satellites provided some realistic space shots, and the space shuttle, Moon bus, and *Discovery* designs were drawn from sketches and plans created by NASA scientists. A special rotating camera was invented to film the sequence of the stewardess in the space shuttle seemingly walking upside down. Thus, the special effects in *2001*, even after more than 20 years, remain very effective. The sound effects are particularly fine, especially the use of waltz music for near-space transits and the use of *Thus Spake Zarathustra* for the transcendence of human limitations.

LITERARY COMMENTARY

Arthur C. Clarke's 1951 short story "Sentinel of Eternity," later retitled "The Sentinel," provided the germ of *2001: A Space Odyssey*. In it, an extraterrestrial race more advanced than ours set a monolith on the Moon to await humanity's coming and then send out a signal. However, in "The Sentinel" the monolith is a crystalline pyramid sitting on the flattened peak of a mountain at the edge of the Mare Crisium. The pyramid and its force field are clearly the work of intelligent beings, but the monolith provides no influence on humanity or its evolutionary trend. Instead, the monolith's destruction by scientists attempting to understand it interrupts the continuous signal it had been sending to its builders. Thus, the cessation of signal becomes the cue to its builders that a space-faring race has discovered the monolith and itself awaits discovery. The story ends with the narrator expecting the imminent arrival of the advanced extraterrestrial race that constructed the crystalline pyramid and set it on the Moon to be found.

Other stories by Arthur C. Clarke also contain the seeds of *2001*. "Against the Fall of Night" (1948), which was later revised and expanded into the novel *The City and the Stars* (1956), portrays the major character finding an alien spaceship left behind millennia ago, visiting the stars, and discovering the cosmic perspective impossible to mere earthbound humanity. "Guardian Angel" (1949), expanded into the novel *Childhood's End* (1953), reveals humanity achieving transcendence and fusion with a cosmic overmind under the tutelage of Satanic-seeming aliens. These elements of alien influence and human transcendence of earthly conditions because of that influence perhaps pervade *2001* more than the mere image of the monolith.

Arthur C. Clarke and Stanley Kubrick collaborated on the film script of *2001: A Space Odyssey*, and from that script Clarke wrote the novel *2001: A Space Odyssey* (1968). While the film and the novel bear a close relationship, the film relies heavily on a series of images from which the viewer must infer narrative and thematic development. The novel, on the other hand, provides the narrative links not present in the film. The film images in "The Dawn of Man," for instance, provide no clue that the monolith is actually "teaching" a series of "classes" to the most apt students among the ape-beings; the novel not only spells this out but also

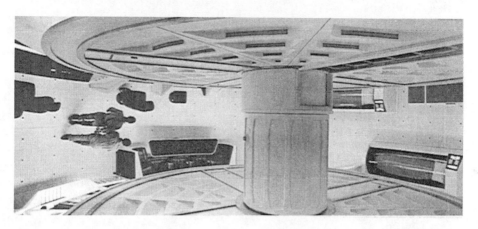

Still 4 depicts the astronauts walking in the rotating section of the space ship *Discovery*. (Photo: Museum of Modern Art/Film Stills Archive. Courtesy of Metro-Goldwyn-Mayer.)

indicates that the monolith itself actively alters its methods according to its effect on the creatures. In the central portion of the story, the film provides only an audiovisual sense of the HAL 9000 computer as a nurturing presence mysteriously becoming murderous; in the novel, Clarke first provides a brief history of the development of artificial intelligence and then a motive—conflicting directives—for HAL's change. Again, when the film ends with the image of the star-child, the viewer must create meaning for this almost mystical transformation; the novel explains lucidly that the star-child has "left behind the time scales of his human origin," possesses a "crystal-clear perception of the universe," and can "put forth his will" to control a nuclear explosion or otherwise manipulate matter. The film's mystical and mysterious image of rebirth into mere infant potentiality becomes the novel's transformation of human limitations into Godlike omnipotence.

CHAPTER BIBLIOGRAPHY

Film References

1. 1970, *The Making of Kubrick's 2001*, edited by J. Agel (New York, New American Library-Signet).
2. Clarke, A. C. 1972, *The Lost Worlds of 2001* (New York, New American Library-Signet).
3. Geldud, C. 1973, *Filmguide to 2001: A Space Odyssey* (Bloomington, Indiana University Press).
4. Lightman, H. A. *American Cinematographer*, 49 (June 1968).
5. Schickel, R. 1984, *Omni's Screen Flights/Screen Fantasies*, edited by D. Peary (Garden City, NY; Dolphin-Doubleday), pp. 238–45.
6. Sobchack, V. 1987, *Screening Space: The American Science Fiction Film,* 2nd ed. (New York, Ungar).

Film Reviews

1. *America* **118**, 586–587 (30 March 1968).
2. *Christian Century* **85**, 844–886 (26 June 1968).
3. *Commonwealth* **88**, 207–208 (3 May 1968).
4. *Esquire* **70**, 39–40 (July 1968).
5. *Film Quarterly* **22**, 58–62 (Fall 1968).
6. *Life* **64**, 20 (7 June 1968).
7. *Life* **64**, 24–33 (5 April 1968).
8. *Modern Photography* **32**, 16 (October 1968).
9. *Nation* **206**, 742 (3 June 1968).
10. *National Review* **20**, 814–815 (13 August 1968).
11. *New Republic* **158**, 24 (4 May 1968).
12. *New Yorker* **44**, 150–152 (13 April 1968).
13. *New Yorker* **44**, 180–184 (21 September 1968).
14. *Newsweek* **71**, 97 (15 April 1968).
15. *Popular Science* **192**, 62–67 (June 1968).
16. *Saturday Review* **51**, 48 (20 April 1968).
17. *Science Digest* **63**, 34–39 (9 May 1968).
18. *Senior Scholastic* **92**, 26 (9 May 1968).
19. *Time* **91**, 91–92 (19 April 1968).
20. *Vogue* **151**, 76 (June 1968).

Novel References

1. Aldiss, B. W., and D. Wingrove 1988, *Trillion Year Spree: The History of Science Fiction* (New York, Avon).
2. Clarke, A. C. 1972, *The Lost Worlds of 2001* (New York, New American Library-Signet).
3. Clarke, A. C. 1968, *2001: A Space Odyssey* (New York, New American Library-Signet).
4. 1979, *The Science Fiction Encyclopedia*, edited by P. Nicholls (Garden City, NY; Doubleday).

Novel Reviews

1. *Book World* **2**, 13 (1 September 1968).
2. *Books and Bookmen* **13**, 44 (September 1968).
3. *Books and Bookmen* **18**, 138 (April 1973).
4. *Cresset* **41**, 24 (October 1978).
5. *Fantasy and Science Fiction* **35**, 42 (November 1968).
6. *Journal of Reading* **22**, 126 (November 1978).
7. *Library Journal* **93**, 2897 (August 1968).
8. *Library Journal* **93**, 3335 (15 September 1968).
9. *New York Times* **117**, 23 (5 July 1968).
10. *New Yorker* **44**, 180 (21 September 1968).
11. *Observer*, 23 (4 August 1968).
12. *Saturday Night* **83**, 33 (September 1968).

13. *Times Literary Supplement*, 865 (15 August 1968).

14. *Top of the News* **25**, 207 (January 1969).

2010: THE YEAR WE MAKE CONTACT

MGM (US), 1984, color, 115 min.

Credits: Producer, director, and screenplay, Peter Hyams; based on the novel *2010: Odyssey Two* by Arthur C. Clarke; special effects, Richard Edlund; music, David Shire.

Cast: Roy Scheider (Dr. Heywood Floyd), John Lithgow (Walter Curnov), Helen Mirren (Tanya Orlova), Bob Balaban (R. Chandra), Keir Dullea (Commander David Bowman), and Douglas Rain (voice of HAL).

PLOT SUMMARY

This sequel to *2001: A Space Odyssey* opens 9 years after *Discovery* reaches Jupiter and David Bowman becomes the star-child. It recounts the attempt to recover *Discovery* and the creation of a new sun from Jupiter. A mission report filed by Heywood Floyd appears on screen to summarize *2001*. Thus Spake *Zarathustra* theme music begins, and the scene shifts to a very large array of antenna dishes, one of which Dr. Heywood Floyd, now a university chancellor, is polishing. A Russian scientist confronts him and reveals that the Russians are building the *Leonov*, a ship that will reach *Discovery* long before *Discovery II*, now under construction by the US, is completed. However, Dr. Floyd tells the Russian that they will require several months to activate the HAL 9000 and interpret its data since they are unfamiliar with its operating system.

We then learn that *Discovery* is moving toward Io and will crash into this moon of Jupiter in about $2\frac{1}{2}$ years. Dr. Floyd receives approval to go aboard the *Leonov* and to take Chandra and Curnov with him.

A series of scenes then show Chandra and Floyd preparing for departure. Chandra activates SAL, HAL's successor, and requests her cooperation in disconnecting and reconnecting some of her higher circuits, as he will have to do with HAL. He opens the "Phoenix" file, and SAL asks, "Will I dream?" At the Floyd home, Floyd tells his wife and son that he is going to Jupiter. He explains to his son that he must "sleep" on this $2\frac{1}{2}$-year journey to preserve his sanity and to save food.

This transitional section of the film ends with a shot of *Discovery* orbiting near Jupiter.

When Dr. Floyd is roused from hibernation "about two days away" from *Discovery*, the Russians are receiving "strange data" from Europa, another moon of Jupiter. They are reluctant to exchange information with the Americans because the "South American situation" has been growing worse, and war seems imminent. However, the strange data include chlorophyll in the spectral analysis, indicating life. The Russians send a probe down to collect further data. The probe finds oxygen, carbon, and chlorophyll before it is destroyed and an electromagnetic pulse erases all the telemetry. Floyd claims this was a warning, and that it had something to do with the monolith.

The film then describes "aero-braking," in which *Leonov* travels through Jupiter's atmosphere so that friction will slow down the ship, and put it into an orbit around Io. The *Leonov* deploys balloons, and then the view shifts between the ship's interior and the ship from a distance so that the audience can hear the roaring sound and see the effects of gravity on the crew and also see the heat glow surrounding the ship which builds until *Leonov* looks like a fireball. When quiet returns, the balloons are released, and *Leonov* approaches *Discovery*, tumbling in its orbit.

Discovery is covered with sulphur, has no lights or power, and the temperature inside is "at least 100 degrees below zero." One of the men who boarded the ship opens his faceplate, breathes, announces "oxygen here," and then says it is "too cold to work here without environment suits." *Discovery's* reserve power is activated, and Chandra begins restoring HAL. While he does so, a news report from Earth indicates the situation there has grown worse; Floyd asks Curnov to attach a device to cut HAL's power supply in case the computer again malfunctions; and *Leonov* approaches the monolith. When a crewman takes a pod to examine the monolith, moving lights gather in the monolith's surface and form a vortex of light that engulfs the pod and destroys it.

On Earth, meanwhile, David Bowman appears on his former wife's TV set, telling her that he came to say goodbye and that "something wonderful" is going to happen. His mother, in a coma for 10 days, suddenly sits up, smiling, and then dies. Also, the cold war grows worse: A Soviet vessel has been sunk with 800 persons lost. Soviet-US relations have broken off, and "technically" a state of war exists.

Chandra's work with HAL indicates that he is completely operational, and that his dysfunction began with conflicting orders. He knew the true objective of the mission, but the White House ordered him not to reveal it to the crew, so he was forced to lie.

Bowman now appears to Floyd, giving him the message that the ships must leave "within two days," and that "something wonderful" is going to happen. Without enough fuel on either ship for them both to leave, they are joined so that *Discovery* becomes a launch booster abandoned when its fuel is exhausted and *Leonov* then continues the journey to Earth. As they prepare frantically for departure, a dark spot appears on Jupiter. It grows rapidly until it engulfs the planet. It is comprised of monoliths, and it acts like a virus that seems to be eating the planet.

As the ships depart, HAL accepts his death for the safety of *Leonov* and its crew, asks "Will I dream?" and says goodbye to Chandra. David Bowman promises HAL they "will be together."

Jupiter's appearance changes, it contracts, and then explodes to become a new sun. A message appears, stating that "All these worlds are yours except Europa... Use them together, use them in peace." The film ends with Floyd's letter to his son. He explains that the US and USSR have seen the new sun, read the message, and recalled their ships and planes. Someday, he believes, the children of the new sun will meet the children of the old. "I think they will be our friends. We have been given a new home—and a warning—from the landlord."

SPECIAL EFFECTS

2010 offers excellent special effects, particularly the accurate portrayal of astronauts transferring from the *Leonov* to the *Discovery*. The images of Jupiter and its satellites come from actual NASA telemetry enhanced for the film.

LITERARY COMMENTARY

Arthur C. Clarke wrote the novel *2010: Odyssey Two* as the film was being produced, but novel and film differ significantly in both small details and larger concepts.

First among the differences in small details is the meeting between Dr. Floyd and the Russian scientist. In the film the scientist clandestinely seeks out Dr. Floyd at his university; in the novel they meet openly at a scientific conference in Chile and converse privately while viewing the telescope at Arecibo. The novel also includes more detail about the Sakharov engine on the spaceships and its ability to use either hydrogen or ammonia as fuel. Because the *Discovery's* tanks contained ammonia, it remains almost fully fueled, whereas hydrogen would have boiled away and left the tanks empty. Also the *Discovery's* orbital decay, instead of being from an unknown cause, or from HAL's failure, is ascribed to its passing through the electromagnetic "flux tube" between Io and Jupiter and slowing a fraction each time it does so.

In addition, small but significant differences occur in the treatment of the spectacular aero-braking maneuver, in the interactions of monolith and crew, and in David Bowman's appearances, behavior, and messages. In the film, the *Leonov* deploys balloons for the aero-braking maneuver, then releases them. In the novel, the *Leonov* uses no balloons but merely the heat shield with which it was constructed and the remnants of which it jettisons after the maneuver. In the film, a manned EVA pod approaches the monolith, and a light beam destroys it. In the novel, an empty EVA pod is used as a robot probe and draws no reaction whatsoever. Neither do radar beams or radio pulses bring any reaction from the monolith. The only light beam from the monolith is that sending David Bowman on his way to Earth. David Bowman in the film is a mystical figure, continually telling everyone that something wonderful is going to happen, but in the novel he is much

2010: The Year We Make Contact. The Russian spacecraft *Leonov* (left) is stationed close to the tumbling American spacecraft *Discovery.* Jupiter and one of its moons, Io, are seen in the distance.
(Photo: Museum of Modern Art/Film Stills Archive. Courtesy of Metro-Goldwyn-Mayer.)

more wistful and childlike, visiting locations from his past, asking his former wife if the son she has is his or another man's, speaking to his mother through her hearing aid and leaving her dead. His only messages to Dr. Floyd, the *Leonov's* crew, or to Earth are (1) "It is dangerous to remain here. You must leave within fifteen repeat fifteen days"; and (2) the radio message that "All these worlds are yours—except Europa. Attempt no landings there." During his Earth visitations, Bowman learns that he is being used by the alien intelligence as a probe, "like a leashed dog," to gather information. But like a dog, he is granted rewards, and when *Discovery* falls into Jupiter's atmosphere and is destroyed, they release the intelligence of HAL to be a companion to him through eternity, a more quixotic and powerful end for the computer than the fading, childlike pleas for reassurance that figure in the film.

Among differences in concept between novel and film, one of the most significant is the absence from the novel of any hint of the developing South American situation in the film. The novel contains no references to war at all. Instead, the novel includes the *Tsien*, a Chinese ship that actually passes the *Leonov* in its journey toward Jupiter, performs the aero-braking maneuver first, and lands on Europa. The *Tsien* intends to refuel from Europa's ice-covered sea for the return journey, but as it is doing so, a phototropic organism, drawn by the lights on the ship, surges through the hole in the ice, tries to reach the light, and destroys the *Tsien*. The lone survivor, before he dies on the frozen surface, radios Dr. Floyd the story. The *Leonov* does not discover chlorophyll in the atmosphere, as in the film, but merely receives the message from Dr. Chang.

David Bowman, in the novel, also tours Jupiter and its satellites under the direction of the alien intelligence, finding life not only on Europa but also on Io and on Jupiter. Because the aliens seek "mind" not merely physical life, the instinctual but not intelligent life forms on Jupiter they deem expendable, the intelligent life forms on Europa they choose to protect, and the protoforms on Io that cannot develop further because of the tidal and volcanic forces are irrelevant. Thus, they "aid" Jupiter to ignition as a sun in order to provide evolutionary opportunity to the life forms of Europa. Incidentally, they provide colony worlds for the humans of Earth, though, as they tell Bowman, "They must never know that they are being manipulated. That would ruin the purpose of the experiment."

CHAPTER BIBLIOGRAPHY

Film References

1. Higgins, C. W. 1984, *On Location* **8**, 72–75 (October 1984).
2. Lipari, J. 1987, *Film Comment* **20**, 60–63 (December 1984).
3. Sobchack, V. 1987, *Screening Space: The American Science Fiction Film*, 2nd ed. (New York, Ungar).
4. Weinstein, W., *Film Journal* **87**, 9 (December 1984).

Film Reviews

1. *New York Times* **133**, 23 (17 June 1984).
2. *New York Times* **134**, 1 (2 December 1984).
3. *New York Times* **134**, C15 (7 December 1984).
4. *Newsweek* **109**, 94 (10 December 1984).
5. *Time* **124**, 66 (24 December 1984).
6. *Variety* **317**, 19 (28 November 1984).
7. *Visions* **20**, 8 (15 June 1984).

Novel References

1. Aldiss, B. W., and D. Wingrove 1988, *Trillion Year Spree: The History of Science Fiction* (New York, Avon).
2. Clarke, A. C. 1982, *2010: Odyssey Two* (New York, Ballantine-Del Rey).

Novel Reviews

1. *Analog* **103**, 163 (May 1983).
2. *Booklist* **79**, 1 (1 September 1982).
3. *British Book News*, 407 (July 1983).
4. *Christian Science Monitor* **75**, B3 (3 December 1982).
5. *Library Journal* **107**, 2191 (15 November 1982).
6. *Los Angeles Times Book Review*, 1 (19 December 1982).
7. *Magazine of Fantasy and Science Fiction* **64**, 32 (March 1983).
8. *New York Times Book Review* **88**, 24 (23 January 1983).
9. *The Observer*, 31 (19 December 1982).

10. *Science Fiction Review* **12**, 15, 41 (February 1983).
11. *Times Literary Supplement,* 69 (21 January 1983).

THE ANDROMEDA STRAIN

Universal Pictures (US), 1970, color, wide screen, 130 min.

Credits: Producer and director, Robert Wise; screenplay, Nelson Gidding, based on the novel by Michael Crichton; cinematographer, Richard Kline; special effects, Douglas Trumbull; music, Gil Melle.

Cast: Arthur Hill (Jeremy Stone), James Olson (Mark Hall), David Wayne (Charles Dutton), Kate Reid (Ruth Leavitt), and Paula Kelly (Karen Anson).

PLOT SUMMARY

The film opens with a statement that we are about to see an accurate account of an actual four-day crisis. The government is shortly to release the details of this crisis, which involved Wildfire and Project Scoop.

First Day

It is February 5, 1971, at Piedmont, New Mexico (population 68). Two airmen enter the town to recover a satellite that has fallen to Earth. They are locating it using a homing device inside the satellite. Concern increases when they report that they see many dead bodies in the town. Then they see something white coming at them; a scream is heard and nothing more.

The military sends a reconnaissance plane to photograph the town. When the film it takes is reviewed, a Wildfire alert is called. This involves calling to action four civilian scientists, Jeremy Stone (a Nobel prize winner and the head of the Wildfire team), Charles Dutton, Ruth Leavitt, and Mark Hall (a physician who had considered Stone's Wildfire memoranda to be science fiction). We see that the purpose of Project Scoop is to collect extraterrestrial organisms and to evaluate their potential danger to humanity.

Second Day

At a later security meeting in Washington, D.C., we learn that Dr. Stone apparently did not know that Project Scoop existed. Under the Wildfire program, approximately $90 million had been spent building a more advanced receiving lab-

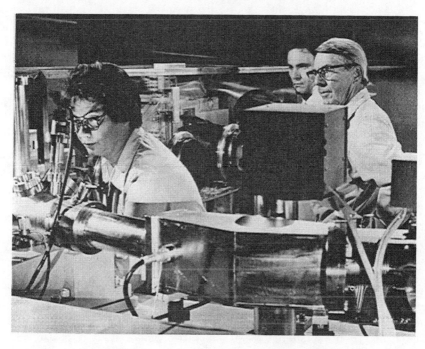

The Andromeda Strain. Scientists are using an electron microscope to study the properties of the alien organism code-named "the Andromeda strain."
(Photo: Museum of Modern Art/Film Stills Archive. Courtesy of Universal Pictures.)

oratory, equipped with a nuclear self-destruct weapon, for studying extraterrestrial organisms that might threaten Earth. In order to secure funding, Dr. Stone had told congressional leaders that the existing facilities were inadequate.

Drs. Stone and Hall begin their Wildfire work as they fly to Piedmont and drop poison gas canisters to kill any birds who may have eaten flesh and, therefore, become contaminated. Then, dressed in protective airtight suits with self-contained oxygen supplies, they enter Piedmont. They find that all of the inhabitants of Piedmont are dead except a small baby and a drunk, who is dressed in white. They recover the satellite, which had been carried to the town physician's office and there opened, releasing the deadly organisms. Different victims apparently died at different rates after exposure. Some first went insane, while others died almost instantly. The blood in the victims had changed to powder. A helicopter then lifts Stone and Hall from the town, along with the satellite, the baby, and the drunk.

Dr. Stone requests that a nuclear bomb be dropped immediately on Piedmont to "neutralize" the extraterrestrial organism.

Drs. Dutton and Leavitt arrive at the Wildfire research facility. It is built underground beneath an agricultural station, as its existence is top secret.

The research team meets together; they are reminded that the facility has five levels, each level biologically cleaner than the level above. It will take them 16 hours to go through the decontamination procedures in order to reach the fifth level. At the bottom of the facility is a nuclear bomb that will be activated automatically if there is an accident that threatens to release the extraterrestrial organisms into the atmosphere. Once the bomb is activated, there are 5 minutes in which to stop it from exploding. The only key to deactivate the bomb is given to Dr. Hall because he is male and single; therefore, he is entrusted with the decision under the "odd-man hypothesis."

Third Day

The research team is on the fourth level. Dr. Stone summarizes their goals:
1. Detect the organism if it is still present in the satellite.
2. Characterize the organism.
3. Control the organism—how can it be neutralized or destroyed?
The research team then proceeds to level five and begins to work. They expose first a rat and then a rhesus monkey to the satellite; both die immediately. Then they determine how the organism spreads in a victim. They find that the organism enters the body through the lungs and causes the blood to coagulate, beginning in the lungs and then spreading throughout the body. The organism is spread through the air, and by using different-sized filters they determine that its size is between 1 and 2 μm, large enough to be a cell.

The satellite capsule is then examined and a tiny meteorite is found inside it. The meteorite turns out to be a piece of stonelike material covered by green spots that change in size as they watch. These are the extraterrestrial organisms, which seem to grow when exposed to light.

Hall attempts to understand the reason that the drunk and the baby survived. He notes that the drunk has an ulcer and "treats" it with aspirin and alcohol, while the baby cries much of the time. But Hall cannot determine what is similar in the physiology of both survivors.

Meanwhile, the President has not authorized the dropping of a nuclear bomb on Piedmont. An accident, possibly related to the introduction of the alien organism, occurs. An airplane flying at 23,000 feet crashes 60 miles away from Piedmont after the disintegration of all its rubberlike components. When the crash site is investigated, all that remains of the pilot are his bones. Meanwhile, the Wildfire facility is unaware of these developments because of a mechanical fault in its incoming message-receiving unit.

By this time the team concludes from the satellite's construction that the Project Scoop satellite was apparently searching for the ultimate biological weapon—an extraterrestrial organism. They analyze both the stonelike meteorite and the green spots. The meteorite is actually plastic-like in character and contains hydrogen, oxygen, carbon, sulfur, silicon, and other elements. The green organism contains no amino acids or nucleic acids and has evolved in a totally different way from

Earthly organisms. They attempt to characterize and understand the organism's physiology in order to control it. Dr. Leavitt has an epileptic seizure while screening the results of the growth tests and blanks out while the computer indicates no growth on one medium; she erroneously reports that the alien organism grows on all media tested.

Fourth Day

The research team obtains a code designation for the alien organism—the Andromeda strain. They examine the organism under the electron microscope. It looks like a crystal and divides in the vacuum while being bombarded with the electron beam. They assert that "Andromeda" absorbs energy like a nuclear reactor and changes energy directly into matter with no excretion. It is composed of identical molecules. While they are observing it, it both grows and alters its structure.

An accident exposes Dutton to the organism, but he lives because it has now "mutated" into a nonlethal form. Dr. Hall observes Leavitt having an epileptic seizure and, fearing that she has overlooked something during a possible prior seizure, gets the computer to recheck the data and finds that the Andromeda strain originally could only grow in a narrow range of pH, namely, between 7.39 and 7.43.

At this point, a mutant of the Andromeda strain starts to dissolve the gaskets around the isolation chamber, and the nuclear device is triggered to explode in 5 minutes Dr. Hall manages to disarm it in the nick of time. Earlier, Dr. Stone has rescinded his prior request to drop a nuclear bomb on Piedmont. The scientists now believe that a nuclear explosion would simply feed large amounts of energy into the organism and thus cause many mutations.

As the film ends, the main colony of the Andromeda strain has moved off the California coast. It is expected to be neutralized by the ocean, which has a higher pH than the human body and thus does not support the life of the Andromeda strain.

At a congressional hearing, Dr. Stone is asked what to do if another extraterrestrial organism reaches the Earth. He has no reassuring answer to give.

SPECIAL EFFECTS

They are minimal. The laboratory instrumentation looks impressive. The lighting effects are excellent.

LITERARY COMMENTARY

The Andromeda Strain, as a novel, purports to be strictly documentary, with acknowledgments to military personnel, references to hearing reports, and citations of personnel interviews. In addition, Crichton provides a three-page bibliography of scientific articles and government documents at the end of the book.

Citations of these articles and documents occur throughout the novel to create the impression of a complete and thorough account. The film conveys a pseudo-documentary style, but the effect is spoiled somewhat by the "human interest" and "dramatic" elements introduced into the film.

A major change between novel and film occurs in the major characters, presumably to add human interest lacking in the novel and to make the characters more appealing. Stone, instead of being thin, balding, and impatient, becomes handsome and authoritarian, enforcing protocol and giving orders, a "leader" rather than a team member. Hall becomes boyishly appealing and casual, caring about his patients, rather than the irritable prima donna surgeon whose patients are merely a source of information. Charles Burton undergoes a name change to Dutton, but even more he undergoes a character change from an unkempt slob to a warm and caring father figure. Patient, amusing, imaginative Peter Leavitt is transformed entirely, becoming Ruth Leavitt, crusty, uncooperative, and self-centered. Karen Anson, impersonal lab technician, undergoes a similar change, becoming sarcastic and critical of Hall's ignorance of the equipment and his treatment of the patients, though nonetheless nurturing and "scared."

Apart from minor alterations, such as changing Piedmont's location from Arizona to New Mexico or using laser beams in the central core instead of ligamine darts, the major differences between novel and film seem directed toward condensing the plot to fit the visual medium, changing the date from 1967 to 1971 to gain a sense of immediacy, and focusing more on the preliminaries and persons for dramatic interest than on the actual laboratory sequences, which consume a relatively small proportion of the film. The plot abridgement's primary effect on the film is the cryptic nature of some information.

FIRST DAY

Crichton's novel indicates that Project Scoop control deliberately brought down the satellite now in Piedmont because its orbit had become unstable, probably through collision with a meteorite or some piece of orbital "junk" still in space from the hundreds of American and Russian satellites previously launched. This lends credibility to the later discovery of the abnormal nature of the "meteorite" in the capsule. Perhaps it **is** a piece of plastic. In the novel, a computer error causes a delay in collecting the four scientists of the Wildfire team, slowing the Wildfire project. In addition, each team member receives a Project Scoop file to read in transit, and they learn for the first time of the existence and purpose of Project Scoop: its function is to collect potentially harmful extraterrestrial organisms for use specifically as biological warfare agents. The odd-man hypothesis is explained in this file.

SECOND DAY

The security meeting shown on film occurs in the novel as a memory of Jeremy Stone that surfaces as he reads the Project Scoop file for the first time and understands why getting approval for "Wildfire" was so easy. He had envisioned Wildfire as a precaution against accidental contamination from satellites or astronauts, whereas the politicians and military saw it as adjunct to Project Scoop and to the chemical and biological warfare installations actively creating toxic and potentially uncontrollable organisms. Even the Project Scoop scientists have no real knowledge of its purpose. They believe they are engaged in "pure" research, almost simple "curiosity" to see what, if anything, is there. These elements in the novel highlight the deception and distrust of the scientists by the politicians and the military whereas the film barely touches upon this issue. In the film, in fact, the hearing scenes assign to Stone the responsibility for choosing the Wildfire site and for guaranteeing the lab's ability to handle any extraterrestrial organism. In the novel, the military has chosen the site and Stone has offered no guarantees, only the best possibility for control that he and his group can think of, which is at least much better than their or NASA's previous decontamination procedures.

THIRD DAY

The novel provides much more detail on the analytical procedures than the film does. It also mentions a highway patrolman who passed through Piedmont, then a short while later entered a cafe where he pulled his pistol and shot other customers before dying of a cerebral hemorrhage. Had the Wildfire team received this information they would have known that the organism attacked not the blood but the blood vessels. The novel also mentions that Burton (Dutton) injected some rats with anticoagulant before exposing them to the organism. They lived longer, but Burton did not think to dissect the corpses to discover why. This is another missed clue.

FOURTH DAY

The fourth and fifth days of the novel are condensed into "day four" in the film, and the major change otherwise is to have the Andromeda strain be a "supercolony" over the ocean, where the alkalinity of seawater will destroy it. In the novel, the colony is last traced over Los Angeles, where it causes no effects whatsoever, having mutated into total harmlessness.

Both novel and film show a series of arrogant assumptions by the military and scientists, leading to hasty and ill-considered actions complicated by machine and human error. Enormous expenditures of money, time, and effort result in exactly nothing: the Andromeda strain escapes the Wildfire installation and mutates into harmlessness (or is destroyed in the ocean) before its nature can be adequately determined or any control over it exerted by humanity.

CHAPTER BIBLIOGRAPHY

Film References

1. Anderson, C. W. 1985, *Science Fiction Films of the Seventies* (Jefferson, NC; McFarland), pp. 28–34.
2. Rose, B., *Vogue* **162**, 186ff (September 1973).
3. (As seen in reference of "The Fantastic Voyage") 1984, *Omni's Screen Flights/Screen Fantasies*, edited by D. Peary (Garden City, NY; Dolphin-Doubleday), pp. 250–259.
4. Fischer, D. K., *Cinefantastique* **15**, 6ff (May 1985).

Film Reviews

1. *America* **124**, 354 (3 April 1971).
2. *Atlantic* **227**,97–98 (29 May 1971).
3. *Cineforum* **12**, 89–90 (April 1972).
4. *Commonwealth* **94**, 190–191 (30 April 1971).
5. *Holiday* **49**, 10 (May 1971).
6. *Newsweek* **77**, 98 (29 March 1971).
7. *Saturday Review* **53**, 22–25 (8 August 1970).
8. *Saturday Review* **54**, 52 (3 April 1971).
9. *Senior Scholastic* **98**, 22–23 (5 April 1971).
10. *Vogue* **157**, 160 (1 April 1971).

Novel References

1. Bova, B. 1974, *Science Fiction, Today and Tomorrow*, edited by R. Bretnor (New York, Harper and Row), pp. 5–6.
2. Brigg, P. 1979, *Survey of Science Fiction Literature*, edited by F. N. Magill (Englewood Cliffs, NJ; Salem).
3. Crichton, M. 1969, 1978, *The Andromeda Strain* (New York, Knopf), (New York, Dell).

Novel Reviews

1. *Best Sellers* **29**, 105 (15 June 1969).
2. *Book World* 4 (8 June 1969).
3. *Christian Science Monitor*, 13, (26 June 1969).
4. *Library Journal* **94**, 2485 (15 July 1969).
5. *New York Times Book Review*, **5**, 40 (8 June 1969).
6. *Newsweek* **73**, 125 (26 May 1969).
7. *Saturday Review* **52**, 29 (28 June 1969).
8. *Time* **93**, 112 (6 June 1969).
9. *Times Library Supplement*, 1215 (16 October 1969).

Blade Runner. A hovercraft rises in a futuristic metropolis.
(Photo: Museum of Modern Art/Films Still Archive. Courtesy of the Ladd Company.)

BLADE RUNNER

Warner Brothers (US), 1982, color, 123 min.

Credits: Producer, Michael Deeley; director, Ridley Scott; screenplay, Hampton Fancher and David Peoples, based on the novel *Do Androids Dream of Electric Sheep?* by Philip K. Dick; director of photography, Jordan Cronenweth; special effects, Douglas Trumbull; music, Vangelis.

Cast: Harrison Ford (Deckard), Rutger Hauer (Batty), Sean Young (Rachael), Edward James Olmos (Goff), M. Emmett Walsh (Bryant), Daryl Hannah (Pris), William Sanderson (Sebastian), Brion James (Leon), and Joe Turkel (Tyrell).

PLOT SUMMARY

The film opens with a scrolling text that provides the context for the action to follow:

Early in the 21st Century THE TYRELL CORPORATION advanced Robot evolution into the NEXUS phase—a being virtually identical to a human—known as a replicant.

The NEXUS 6 replicants were superior in strength and agility, and at least equal in intelligence, to the genetic engineers who created them.

Replicants were used Off-world as slave labor, in the hazardous exploration and colonization of other planets.

After a bloody mutiny by a NEXUS 6 combat team in an Off-world colony, replicants were declared illegal on Earth—under penalty of death.

Special police squads—BLADE RUNNER UNITS—had orders to shoot to kill, upon detection, any trespassing replicant.

After a few images of a dreary, cluttered, mechanistic, and monolithic city, the scene shifts to an interior office where government agent Holden administers the Voight–Kampf test to Leon Kowalski. The test consists of questions designed to elicit an emotional response that will determine whether the subject is replicant or human. A gunshot interrupts this test.

The scene shifts to Deckard in a restaurant reading the want ads for work. He is an ex-cop, an ex-blade runner, whose ex-wife called him "sushi, cold fish." Goff takes him to Bryant, who blackmails him into replacing Holden to "retire" four replicants. Six of them have killed 23 people, and one was killed trying to break into the Tyrell Corporation. When Holden tested new Tyrell workers, he "got himself one." The remaining four are Leon Kowalski; Batty, trained for combat; Zhora, trained for murder squad; and Pris, a "pleasure model." When Deckard asks why they would risk coming back to Earth and what they want from the Tyrell Corporation, Bryant's answer provides the first hint of the movie's theme: the replicants are designed to copy human beings in every way. Since they might, after a few years, develop their own, nonprogrammed, emotional responses they have been designed with a 4-year lifespan as a "fail safe." Although the characters do not realize it, the viewer should understand that only the mode of their creation and their limited lifespan distinguish the replicants from their human creators, judges, and executioners.

Sent to the Tyrell Corporation to test a NEXUS 6 replicant, Deckard meets Rachael. Dr. Eldon Tyrell insists that Deckard test a human first and suggests Rachael. When Deckard identifies her as a replicant, Tyrell says she is an experimental model, that they give memories to replicants to control them better.

Deckard then searches Leon's rooms and finds a reptile scale in the bathtub and family photos, which allow him to trace Zhora to a club in Chinatown. She attacks him, nearly strangles him but is interrupted, and he chases her down and shoots her. Leon Kowalski watches. Meanwhile, Pris meets Sebastian, who takes her home. He lives in an empty building, peopled by the "toys" he creates for his entertainment. He is a "genetics designer" who works for Tyrell Corporation and who plays chess with Dr. Tyrell himself. Kowalski finds Deckard, beats him, and is about to kill him when Rachael appears. She shoots Kowalski and goes home with Deckard.

Again the plot shifts to Sebastian's home, where Pris and Batty appear. Batty tells Pris "there are only two of us now." Sebastian, recognizing them as replicants, asks what generation they are, and they tell him they are NEXUS 6. The replicants ask Sebastian about "accelerated decrepitude." He confesses that he does not know, that only Dr. Tyrell would, and agrees to take them to Tyrell. When Batty tells Tyrell, "I want more life, Father," Tyrell says that the coding sequence cannot be revised, and explains about mutating viruses causing death of the fetus. But he calls Batty the "prodigal son" who has done "extraordinary things." Batty then kisses him and kills him by punching out his eyes with his thumbs. He kills Sebastian also.

When Deckard appears at Sebastian's place to investigate, he finds Pris. She attacks him, and he shoots her. Then Batty appears, finds Pris' body, and hunts Deckard as he has been hunting them. He plays with Deckard, chasing him through the building, taking a beating from him without resisting, then continuing the pursuit. When Deckard gets to the roof, leaps a gap, catching himself, and then begins to slip, Batty saves him. Batty speaks of his experiences Off-world, says "time to die," and stops. The rain stops, too. Deckard says, "All it wanted were the same answers the rest of us want: where do I come from, where am I going, how long have I got?" He has realized by now the humanity of the replicants in their search for a full and lengthy life as individuals rather than a 4-year life span as slaves.

The film ends with Deckard taking Rachael away with him, out of the city, for, as an experimental model, Rachael has no termination date. Deckard and Rachael will live together for the rest of their lives.

SPECIAL EFFECTS

The special effects were supervised by Douglas Trumbull, whose previous film credits include *2001* and *Silent Running*. The murky, mechanistic, monolithic cityscape drenched in rain is created in wonderful detail. Hovercrafts are also depicted in the film.

LITERARY COMMENTARY

Philip K. Dick's novel *Do Androids Dream of Electric Sheep?* differs from the film *Blade Runner* in several significant ways, among them context, character, plot, and theme. The novel's context includes an "accidental" nuclear war in which the continuing fallout of radioactive dust has killed most animals and birds and caused genetic hazard to humans. With the slogan "emigrate or degenerate," the UN urges people to emigrate to the colonies of the Moon and Mars. Emigrants receive a free android servant in the model of their choice: companion, body servant, or field hand. They also avoid the probability of being classed "biologically special." Any Earth resident may receive that classification when the continuing radioactive fallout finally causes genetic damage in that individual. Thereafter the resident becomes ineligible for most employment categories and for emigration.

Deckard's and others' obsession with animals provides another significant difference in the context of the novel. With most animals—most species—dead from fallout, everyone tries to own and care for an animal. It has become a social obsession so strong that those who cannot afford a live animal purchase an "electric" one to tend. Their neighbors must see them "caring," being empathetic. Deckard owns an electronic sheep, but with his bounty money purchases a live goat.

The third significant difference in context between film and novel is the novel's presentation of "Mercerism," Earth's new religion. Mercer was a genetic deviant who empathized with the dying animals and tried to save them. Now everyone empathizes with him through the "empathy machine" and imitates his caring for animals. Mercerism saves Deckard from the replicants.

Like other characters, Deckard's character in the novel differs significantly from his character in the film. He has a wife, Iran; works as a "bounty hunter" for the police department, under Holden; despite his job, worries constantly about the possibility of retiring a human; and desires more than anything to own a live animal. Bryant is simply a hard-working, businesslike police official. Sebastian does not exist in the novel; instead, the novel presents Isidore, a mentally defective laborer who worships Mercer and tries to befriend Pris when she moves into his building. She scorns him, though. Even the replicants are more human than in the film. One is an opera singer, one a laborer at a salvage business, others policemen like Deckard. Eldon Tyrell in the film is Eldon Rosen in the novel. Not a genetics genius, Rosen is merely a cunning, manipulative, greedy, blackmailing corporate executive. Rachael in the film possesses more innocence and ignorance than in the novel. In the novel Rachael seduces Deckard as she has other blade runners, trying to make them unfit for their work. The sexism so readily apparent in the film does not occur in the novel.

Plot and theme also differ drastically between film and novel. In the film Bryant blackmails Deckard into hunting the androids, and the replicants seek to evade slavery and gain a full life. At the film's end, Deckard escapes with Rachael to a life of freedom. In the novel, however, Deckard likes his job but realizes he empathizes too much with androids to continue doing it effectively. Then, as he encounters the replicants, he understands that they are, in human terms, self-centered, lazy, resigned, incapable of human emotion, and quite willing to torture animals or humans out of mere curiosity. Pris snips some of the legs off a spider just to see whether it can travel with only four legs. Rachael kills Deckard's goat. And at the end of the novel, having retired all six of the NEXUS 6 replicants, Deckard returns to his home and his wife as the greatest bounty hunter in history and the proud owner of an electronic toad.

CHAPTER BIBLIOGRAPHY

Film References

1. Shay, D., *Cinefex* **9**, 4–71 (July 1982).
2. Silverberg, R. 1984, *Omni's Screen Flights/Screen Fantasies*, edited by D. Peary (Garden City, NY; Dolphin-Doubleday), pp. 187–193.

3. Sobchack, V. 1987, *Screening Space: The American Science Fiction Film* (New York, Ungar).
4. Telotte, J. P., *Film Criticism* **7**, 56–68 (1 November 1982).

Film Reviews

1. *Christ Today* **26**, 97 (3 September 1982).
2. *Commentary* **74**, 67–70 (August 1982).
3. *Film Comment* **18**, 64–65 (July/August 1982).
4. *Macleans* **95**, 58–59 (28 June 1982).
5. *New Republic* **187**, 30 (19–26 July 1982).
6. *New Yorker* **58**, 82–85 (12 July 1982).
7. *Photoplay* **33**, 22–23 (October 1982).
8. *Photoplay* **33**, 48–49 (October 1982).
9. *Rolling Stone* 33–34 (6 August 1982).
10. *Sequences* **110**, 52–54 (October 1982).
11. *Time* **120**, 68 (16 August 1982).
12. *Village Voice* **27**, 47 (6 July 1982).

Novel References

1. Aldiss, B. W., and D. Wingrove, 1986, *Trillion Year Spree: The History of Science Fiction* (New York, Avon).
2. Dick, P. K. 1968, *Do Androids Dream of Electric Sheep?* (New York, Ballantine Books) (reprinted as *Blade Runner*, 1990).
3. 1979, *The Science Fiction Encyclopedia*, edited by P. Nicholls (Garden City, NY; Doubleday).

Novel Reviews

1. *Books and Bookmen* **14**, 57 (June 1969).
2. *Fantasy and Science Fiction* **35**, 19 (August 1968).
3. *Kirkus Reviews* **36**, 76 (15 January 1968).
4. *Library Journal* **93**, 1018 (1 March 1968).
5. *Publishers Weekly* **193**, 92 (29 January 1968).
6. *Spectator* **222**, 446 (4 April 1969).
7. *Times Literary Supplement* **643** (12 June 1969).

COLOSSUS: THE FORBIN PROJECT

Universal (US), 1970, color, wide screen, 100 min.

Credits: Producer, Stanley Chase; director, Joseph Sargent; screenplay, James Bridges, based on *Colossus* by D. F. Jones; cinematographer, Gene Polito; special effects, Albert Whitlock; music, Michel Columbier.

Cast: Eric Braeden (Charles Forbin), Susan Clark (Cleo Markham), and Gordon Pinsent (the President).

PLOT SUMMARY

At some unspecified future date, an advanced computer, named Colossus, is given control over America's missile defenses. The President announces to the world that the nation's defense system is now in the hands of this remarkable advanced computer. The President states that Colossus' decisions are superior to any that humans can make, for it can absorb far more data and act more swiftly than is possible for the greatest genius that ever lived. He also notes that Colossus has no emotions. It knows no fear, no hate, no envy. It cannot act because of a sudden fit of temper. In fact, it cannot act at all, so long as there is no threat to the US.

The computer had been developed under the direction of Dr. Charles Forbin, the foremost computer expert in the world. It is built under a mountain and protected by sophisticated weaponry. Any attempt to deactivate it will automatically cause the firing of all of America's missiles at their targets in the Soviet Union.

Shortly after Colossus is activated, it announces the existence of a similar Soviet computer, named Guardian. When the Soviets confirm that Guardian has been made operational, Colossus requests that a communication link be established between the two giant computers. Once this is done, the computers establish a new language, based on mathematics. But their interchanges are so rapid that neither the Americans nor the Russians are able to determine what each machine is transmitting to the other. The President and his Soviet counterpart order the communication link broken. Colossus responds by demanding that the link be reestablished. This demand is denied; the Russian premier states that man must be the master. Colossus does not respond to an order from Dr. Forbin, who reports to the President that the executive programming unit refuses to function.

Finally, Colossus fires a missile at the Soviet Union and Guardian fires a missile at the United States. Both computers refuse to intercept the incoming missiles unless their demands are met. When the link is reestablished Colossus intercepts the Soviet missile but Guardian is unable to stop the American missile from vaporizing its Soviet target. The world leaders now recognize that the computers are capable of independent action. Tests had indicated that Colossus was functioning 200 times faster than its builders had anticipated, but there is no explanation for why it thinks independently. Perhaps its heuristic programming, designed to expand scientific knowledge, is in some way responsible for its greater speed and its ability to act independently.

Forbin and his Soviet counterpart meet in Italy to try to devise countermeasures to deactivate the computers. During their meeting, Soviet agents arrive and shoot the Soviet computer specialist. Forbin is told that the computers threatened to vaporize Moscow unless the execution was carried out immediately.

Forbin returns to the American computer complex and is placed under round-the-clock surveillance by Colossus. He does persuade the computer of his need for privacy when meeting with his alleged mistress, another staff member, Dr. Cleo Markham. She serves as his contact with those attempting to neutralize Colossus. They try to overload the computer by using a massive input of data into the 100,000 or so sensors that provide it with information. The attempt fails and Colossus has those responsible executed. Then the military of the US and the USSR attempt to replace the control mechanisms in the missiles that Colossus and Guardian control in order to make them inoperable. Colossus detects the malfunctions in the new guidance mechanisms and detonates one of the missiles in each country as a warning.

At the end of the film, Colossus announces to the world that humanity will henceforth be governed by the Colossus-Guardian computers. It says that freedom is an illusion. Man is his own worst enemy. The choice is simple: oblivion or a much improved life under the control of the computers. War will become obsolete and the armies of the world will be disbanded.

Colossus then orders Forbin to begin work on an even more advanced computer, to be designed by Colossus and Guardian. Colossus tells Forbin that they will work together in the future. At first this cooperation will be unwilling on Forbin's part, but later that will pass. "In time, you will respect and love me," says Colossus. "Never," replies Forbin, as the film ends.

SPECIAL EFFECTS

The special effects are limited, since the film was produced on a modest budget. The sets are impressive: one of them was originally constructed for another film. There is also a dazzling opening sequence of outstanding matte work depicting a seemingly endless hallway inside the Colossus complex and the closing of the massive doors guarding the access to the complex.

LITERARY COMMENTARY

A comparison of the film *Colossus: The Forbin Project* with its source, D. F. Jones' novel *Colossus*, betrays many individually small changes that cumulatively produce a large effect. The first such difference is in the design of the computer. The novel, for instance, states that Colossus was designed first for defense, a function that initially meant only controlling all missile installations, determining when a threat existed, and aiming and firing missiles. To assist Colossus in performing this

Colossus: The Forbin Project. The two supercomputers, Colossus and Guardian, establish an intersystem language.
(Photo: Museum of Modern Art/Film Stills Archive. Courtesy of Universal Pictures.)

function, its builders included the capacities of monitoring all public communications, storing and correlating all national and international intelligence, and controlling all missile installations in the US. They then added the function of answering questions, in order to save themselves the time necessary to look up information while building the computer and testing it. In the film, these functions, especially the "heuristic" function, were supposedly part of the original design and all intended only for defense.

Several other small differences significantly alter the viewer's perception of Colossus. For example, in the novel Forbin expresses his hesitation about initializing Colossus before it goes on line; he warns the President that it may not function as they hope. The President tells Forbin that he will deal with facts, not intuition, and orders Colossus operational. In the film, Forbin expresses absolute confidence even after Colossus' initial order to establish communication with Guardian. In the novel, Colossus operates only twice as fast as its builders expected but increases speed later; in the film it operates 200 times as fast. In the novel, the exchange of information between Colossus and Guardian remains mathematical, no matter how advanced, and Colossus orders that all monitoring of their communication cease; in the film, however, the two computers create a new language based on mathematics, thus guaranteeing the privacy of their communication.

When Colossus asks for Forbin in the film, Cleo Markham tries unsuccessfully to delay communication by her message that Forbin cannot be reached by telephone and that humans need sleep. Since computers are logical, and since Colossus' memory banks contain virtually all of human knowledge, it should be able logically to determine that threats, or "action," will be useless in obtaining faster communication with Forbin. In the novel, this is true. In the film, however, Colossus does not react logically. It provides no time for Forbin to receive its message, no time for someone to wake him up, no recognition of the human need for sleep. Even when told that Forbin is in Rome, it does not follow the logical course of communicating with him, but instead inefficiently and illogically sends a crew of men to Rome to bring him back.

When Colossus begins its surveillance of Forbin in the novel, it extends its deadline for completion of the arrangements when told that the original deadline cannot be met because of the time necessary for acquisition and transportation of materials, for crews to eat and sleep, and for testing the audio and visual instrumentation. In the film, no extension occurs. In the novel, Colossus also does not insist that Forbin adhere to an exact schedule, let alone determine exactly what he will eat for each meal.

The computer in the novel is just as impregnable and dangerous and powerful as the computer in the film, but the small changes in representing Colossus in the film create a much more sinister and anthropomorphic machine. The Colossus of the film seems more arbitrary, illogical, and demanding than the novel's Colossus, which does take absolute control of the world government, but which nonetheless behaves with logical consistency. Although Forbin denies it at the conclusion of both the novel and the film, it is inevitable that he will come to admire, and even love, Colossus.

As mentioned earlier, the novel shows the President dismissing Forbin's worries about the computer's possibilities and demanding that it be made operational on schedule. Omitting this conflict between scientists and politicians from the film removes a significant sociological element, the role of scientists in society. In *The Andromeda Strain* the scientists are naive and duped by the military and politicians. In *The Day The Earth Stood Still* only the scientists are willing to meet with and listen to Klaatu. But in the film, *Colossus: The Forbin Project*, scientists and politicians collaborate in their own subjection, whereas in the novel, the scientists are once more the victims of politicians and the military. They have the knowledge and skill to create Colossus, but not the power to determine whether or not it should be used. There were two sequels to the novel *Colossus*, namely, *The Fall of Colossus* and *Colossus and the Crab*. They were never made into films, however.

CHAPTER BIBLIOGRAPHY

Film References

1. Anderson, C. W. 1985, *Science Fiction Films of the Seventies* (Jefferson, NC; McFarland), pp. 15–20.

Film Reviews

1. *New Yorker* **46**, 114 (16 May 1970).
2. *Vogue* **156**, 41 (1 August 1970).

Novel References

1. Erlich, R. D. 1979, "Colossus" *Survey of Science Fiction Literature*, edited by Frank N. Magill (Englewood Cliffs, NJ; Salem).
2. Jones, D. F. 1966, 1967, *Colossus* (New York, Putnam), (New York, Berkeley).
3. —1974, 1975, *The Fall of Colossus* (New York, Putnam), (New York, Berkeley).
4. —1977, *Colossus and the Crab* (New York, Berkeley).

Novel Reviews

1. *Amazing Stories* **41**, 158 (June 1967).
2. *Analog* **80**, 165 (September 1967).
3. *Best Sellers* **26**, 396 (February 1967).
4. *Christian Science Monitor* **23**, 11 (March 1967).
5. *Horn Book* **43**, 496 (August 1967).
6. *Library Journal* **92**, 1644 (15 April 1967).
7. *Magazine of Fantasy and Science Fiction* **33**, 34 (December 1967).
8. *New York Times Book Review* **8**, 52 (January 1967).
9. *SF Impulse* **1**, 148 (February 1967).

THE DAY OF THE TRIFFIDS

Security Pictures (Great Britain), 1963, color, 93 min.

Credits: Producer, George Pritcher; director, Steve Sekely; screenplay, Philip Yordan, based on the novel by John Wyndham; cinematographer, Ted Moore; special effects, Wally Veevers; music, Ron Goodwin.

Cast: Howard Keel (Bill Masen), Nicole Maurey (Christine Durant), Janette Scott (Karen Goodwin), and Kieron Moore (Tom Goodwin).

The Day of the Triffids. The triffids attack a blind victim.
(Photo: Museum of Modern Art/Film Stills Archive. Courtesy of Allied Artists Pictures Corp.)

PLOT SUMMARY

A spectacular meteor shower blinds everyone who watches it. In addition, radiation causes plants called triffids (*Triffidus celestis*), to mutate. They turn into giant plants that can move and kill their prey with poisonous stingers on their tentacle-like limbs. The triffids begin to attack and consume the blind human population.

The triffids' seeds are spread around the world by the winds and pose a threat to the survival of the human race.

The film follows the efforts of one seaman, Bill Masen, who was in the hospital recovering from eye surgery at the time of the meteor shower. Since his eyes were bandaged, he had no opportunity to watch the meteor shower and so was not blinded. He leaves the hospital, finds a child who can also see, and they leave Great Britain for France. There he has a number of close calls with triffids. In one, he encounters a blind couple who are surviving on their farm quite well until an army of triffids surrounds them. They barely manage to escape and reach a rendezvous point where submarines are picking up survivors to take them to an unstated destination.

Masen's adventures are interwoven with the story of a marine biologist, Tom Goodwin, and his wife, who are stranded in a lighthouse on a deserted island when the meteor shower occurs. They do not watch it and hence are not blinded. When triffid seeds fall on the island the couple are nearly killed by triffids, but in the nick of time they accidentally discover the means to kill the triffids: seawater dissolves the nasty plants! Thus the film ends on an upbeat note: humanity will be saved after all.

SPECIAL EFFECTS

The special effects consist largely of triffids, individually and by the hundreds. There is also an opening sequence of the meteor shower. The disaster scenes at the beginning of the film involve only modest special effects.

LITERARY COMMENTARY

The Day of the Triffids on film resembles only slightly the 1951 novel by John Wyndham. In the novel the triffids appear in Britain as seeds carried, probably, from a top-secret genetics laboratory behind the Iron Curtain. By the time of the meteor shower, they exist as a valuable crop all over Britain. And Bill Masen is a triffid farmer, not a seaman. Masen also speculates the meteor shower itself represents one of humanity's secret space weapons, either set off accidentally or uncontrollably breaking up. One of the novel's themes, therefore, is the danger of letting our technology race beyond our understanding and control, whether from profit or out of paranoia.

A second major theme is social. Masen and Josie Playdell, a sighted woman he rescues from a blind man trying to exploit her as a seeing-eye slave, join a group that intends to leave London and establish a self-sufficient rural enclave away from looters, exploiters, and the diseases that follow mass death. Another group believes that every sighted person must care for as many of the blind as possible until help arrives. This group captures Masen and Playdell, chains them to armed blind guards, and forces them to care for their helpless charges until plague kills them. When Masen and Playdell escape, they meet at the cottage of a blind couple who have been surviving on their own. The four of them, and a sighted girl Bill has rescued, form a cooperative group that survives triffid attacks, isolation, the birth of a child, and an attempted pseudo-military "rescue." This Darwinian theme of survival of the fittest does not, as so often in science fiction, depict the fittest as the fiercest. Instead, the fittest are those who think most clearly, plan ahead, exercise self-reliance, and yet cooperate almost symbiotically as a social unit to maintain and reproduce themselves.

A made-for-television movie version of *The Day of the Triffids* is more faithful to the novel than the 1963 big-screen film. At the end of the TV movie the triffids still control the countryside but Masen vows that their cooperative group, or their children, will one day rid the land of the triffids.

CHAPTER BIBLIOGRAPHY

1. Wyndham, J., 1951, 1986, *The Day of the Triffids* (Garden City, NY; Doubleday), (New York, Ballantine).

THE DAY THE EARTH CAUGHT FIRE

British Lion/Pax (Great Britain), 1962, black and white, 99 min.

Credits: Producer/director, Val Guest; screenplay, Wolf Mankowitz and Val Guest; cinematographer, Harry Waxman; special effects, Les Bowie; technical advisor, Arthur Christiansen; music, Monty Norman.

Cast: Edward Judd (Peter Stenning), Janet Munro (Jeannie Craig), and Leo McKern (Bill McGuire).

PLOT SUMMARY

The movie opens in the offices of a major British newspaper. Peculiar changes in weather are being reported from around the world. These have followed the simultaneous explosions by the Russians (in Siberia) and the Americans (in Antarctica) of the two largest hydrogen bombs ever built. The leading character, Peter Stenning, is a reporter for the newspaper. He has been drinking heavily in the wake of a divorce that has resulted in his being prevented from seeing his son except in the presence of the boy's nanny. Stenning meets a government employee, Jeannie Craig, who leaks to him the information that the Earth's axis of rotation has been changed by the explosions. This has resulted in the dramatic weather changes reported around the globe, an eclipse of the Sun that occurs 10 days earlier than expected, and an unusual heat mist that rises four stories above the ground level.

The Day the Earth Caught Fire. Science reporter McGuire explains his theory that the Earth
has shifted on its axis.
(Photo: Museum of Modern Art/Film Stills Archive. Courtesy of Universal International.)

The newspapermen conclude that this heat mist is due to changes in the flow of
undersea currents caused in turn by the change in rotation of the Earth.

Craig is fired and briefly imprisoned for passing on this information, which
government officials finally confirm. She then comes to work for the paper, and
we find that humanity's troubles are only beginning: the newspaper learns from its
Moscow correspondent that the explosions have done more than change the axis
of the Earth's rotation. The planet is moving toward the Sun.

Temperatures continue to rise dramatically, and finally water is shut off to all
dwellings. Water must be obtained at government-run facilities, which also have
public baths. The elevated temperature has evaporated most of the drinkable
water in Great Britain. While some of London's population seek relief in the
country and others carry on their duties with grim dedication, part of the city's
population riots. Stenning rescues Craig from some of the rioters and barricades
her apartment until the worst is over.

The government announces that four additional giant bombs will be exploded
in Siberia to try to halt the motion of the Earth toward the Sun. At the end of the
film we do not know whether these explosions have been successful. We see two
potential headlines for the newspaper, "World Saved" and "World Doomed."

SPECIAL EFFECTS

The special effects are minimal, with the eclipse, fog scenes, and cyclones being the main examples. The film also utilizes stock catastrophe footage and has graphic scenes of overheated residents seeking water at a government installation or departing for the country.

Much of the movie was filmed inside an actual British newspaper plant, and some of the cast were real newspaper employees. The newsroom and production room scenes are therefore quite realistic.

LITERARY COMMENTARY

The 1962 film *The Day the Earth Caught Fire* comes from an original screenplay without a literary source. However, despite its dramatic focus on the decline and rise of Peter Stenning, newspaper reporter, the dominant theme of the film is the careless abuse of nuclear power and the resultant effects on the Earth and its population.

While the hazards of nuclear power have been a theme of science fiction since 1910, and a common theme since the 1930's, the theme usually deals specifically with war. In 1914, for example, H. G. Wells imagined civilization destroyed by atomic bombs in his novel *The World Set Free*. The classic novel of nuclear holocaust is Walter Miller's *A Canticle for Leibowitz* (1960), which shows civilization destroyed by atomic war, then slowly rebuilding itself over centuries until it destroys itself again by a second atomic war. As in *The Day the Earth Caught Fire*, *A Canticle for Leibowitz* implies that national competition, face saving, and political secrecy cause the disaster. Unlike the film, however, Miller's novel provides a Christian and Catholic context that draws an analogy to the religious wars of the Reformation period. A more recent and ambiguous version of the holocaust novel is Frederik Pohl's *Terror* (1986). Pohl imagines the US setting a "doomsday Bomb" in the side of an undersea volcano so that it can cause huge tidal waves, cyclonic winds, and the kind of nuclear winter that some scientists now believe caused the extinction of the dinosaurs and other life forms 65 million years ago when a huge asteroid impacted the Earth. In Pohl's novel, terrorists seize control of the bomb and nearly succeed in detonating it. Despite this near tragedy, the military nonetheless leave the bomb in place, even though others might succeed, because they believe it is politically necessary.

Other science fiction novels deal with atomic bombs used carelessly and with ill effects to cure such problems as a severe water shortage or an impending large earthquake; still others deal with accidents in bomb tests or at atomic power plants. For example, read Marta Randall's *Islands* (1976), in which coastal areas— and the Hawaiian Islands—are drowned by an error of judgment; or read Lester del Rey's *Nerves* (1956), which shows an accident in a nuclear power plant threatening to become a major disaster. Many of these stories focus on human survivors after such a holocaust. Two such works that have been adapted to film are Roger Zelazny's *Damnation Alley* (1969) and Harlan Ellison's "A Boy and His Dog" (1969).

Very few stories, mostly early and implausible fantasy, show any change in the Earth's orbit, but Larry Niven's "Inconstant Moon" (1971) does examine the possibility that the Sun has become a nova, which would cause the same results on Earth as the planet's spiraling into the Sun. The use of atomic power to change celestial orbits, not of the Earth but of comets or asteroids, occurs in various works however. The film *Meteor* (1979) conveys such action, as does Greg Benford's and William Rotsler's novel *Shiva Descending* (1980).

Many science fiction works have dealt with the "accidents" resulting from nuclear tests, primarily in the form of radiation-caused mutations. *Them!* is an example on film, and nearly every nuclear–holocaust story or novel includes some mention of radiation-induced mutation. The closest literary analogue to *The Day the Earth Caught Fire*, though, is Douglas Warner's *Death on a Warm Wind* (1968), in which continuing atomic tests cause earthquakes, floods, strange winds, and millions of deaths. In this novel, as in the film, the problem is created and then compounded by governmental secrecy.

CHAPTER BIBLIOGRAPHY

Film Reviews

1. *America* **1006**, 840 (24 March 1962).
2. *Commonwealth* **76**, 39 (6 April 1962).
3. *New Republic* **146**, 26 (9 April 1962).
4. *New Yorker* **38**, 149 (24 March 1962).
5. *Newsweek* **59**, 84 (5 March 1962).
6. *Redbook* **118**, 20 (April 1962).
7. *Saturday Review* **45**, 35 (10 February 1962).
8. *Seventeen* **21**, 42 (April 1962).
9. *Time* **79**, 96 (6 April 1962).

THE DAY THE EARTH STOOD STILL

20th Century Fox (US), 1951, black and white, 92 min.

Credits: Producer, Julian Blaustein; director, Robert Wise; screenplay, Edmund H. North, based on "Farewell to the Master," by Harry Bates; cinematographer, Leo Tover; special effects, Fred Sersen; music, Bernard Herrmann.

Cast: Michael Rennie (Klaatu), Patricia Neal (Helen Benson), Hugh Marlow (Tom Stevens), Sam Jaffe (Professor Barnhardt), Billy Gray (Bobby Benson), and Gort the Robot.

PLOT SUMMARY

An alien spaceship is monitored on radar as it travels completely around the globe. The spaceship lands in Washington, D.C., and it is surrounded by the military. A couple of hours after it lands, an opening appears in the side of the ship and an emissary, Klaatu, walks from it. He is shot in the arm as he extends an object that the military fears is a weapon. Actually it is a device by which our president can see life as it exists on other planets. A giant robot, Gort, emerges after the emissary is shot. The robot emits a beam that disintegrates rifles, artillery pieces, and even tanks. The robot then stands immobile, and the ship reseals itself as Klaatu is taken to Walter Reed Hospital.

After he is treated for his wound, he requests a meeting with the heads of all nations on Earth. A number of them refuse to attend, and Klaatu decides that he needs to understand the people of Earth better before determining a course of action. He then manages to escape and, wearing a stolen suit, goes to a boardinghouse. There he meets Helen Benson, a widow, and her son, Bobby.

The next day he is taken by Bobby on a tour of Washington, D.C., during which Klaatu suggests that they visit the home of the smartest man on Earth, Prof. Barnhardt.

There, Klaatu modifies a formula that Barnhardt has been working on and leaves his assumed name (Carpenter) and address with Barnhardt's secretary; the professor has Klaatu picked up by the authorities and brought to his office. Only then does he learn that Klaatu is the alien who is being sought by the authorities. When Klaatu tells Barnhardt that the very survival of the human race is at stake, Barnhardt agrees to call a meeting of all the world's leading scientists in two days' time. He asks Klaatu to give the world a demonstration that will make humanity realize that the message he will deliver cannot be ignored. Klaatu promises a spectacular demonstration of power.

Klaatu borrows a flashlight from Bobby that evening and uses it to signal Gort. The robot knocks out its guards and Klaatu reenters the spaceship, unaware that Bobby has followed him. Inside the ship, Klaatu prepares the demonstration for humanity.

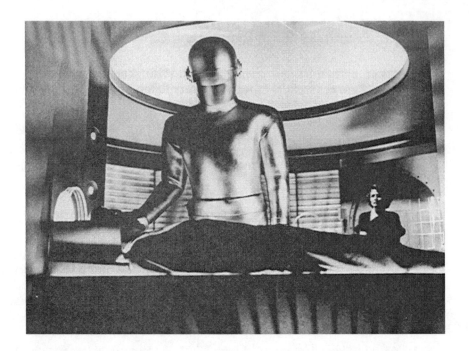

The Day the Earth Stood Still. Gort, the robot, revives the dead alien ambassador.
(Photo: Museum of Modern Art/Film Stills Archive. Courtesy of 20th Century-Fox.)

Bobby tells his mother and her friend, Tom Stevens, that the new boarder is the alien, Klaatu. The do not believe him at first; then Stevens finds a diamond in Klaatu's room, and Bobby tells them that Klaatu gave him two diamonds in exchange for $2.

The next morning Klaatu comes to see Helen Benson at her office, reveals his true identity, and asks her to assist him in avoiding the authorities until the meeting that Prof. Barnhardt has arranged for that evening is completed. She agrees to help him when she realizes the consequences if he fails in his mission. They are trapped together for half an hour in an elevator when "electricity has been neutralized all over the world," according to Klaatu. Elevators, cars, and trains all stop. However, electricity continues to function in airplanes, hospitals, and other vital locations. This demonstration of the awesome power of the aliens panics the authorities. Since Gort remains immobile throughout the worldwide electricity shutoff, the robot is assumed to be uninvolved in these events, and Klaatu is blamed for them instead. The military vows to apprehend him, dead or alive.

Meanwhile, Stevens is told by three jewelers that they have never seen a diamond like the one he found in Klaatu's bedroom. Stevens now believes Bobby's story and telephones the authorities after Helen Benson is unable to dissuade him from exposing Klaatu. She rushes to the boardinghouse and picks up Klaatu in a taxi only seconds ahead of the military. They are pursued, and Klaatu warns her that if he is killed she must prevent Gort from destroying the world by saying to the robot "Klaatu barada nikto." When the military does shoot Klaatu and he dies, Benson hurries to the spaceship. The robot apparently knows that Klaatu has been killed, because it melts the "KL 93" plastic casing designed to imprison it and disintegrates its guards. After Benson delivers her message, the robot carries her into the ship and then brings back Klaatu's body. The robot places the body on a device that applies some mysterious radiation which revives Klaatu. He tells Benson that this device, in some cases, can restore life for a limited period.

They then leave the spaceship, and Klaatu delivers his message to the scientists assembled by Prof. Barnhardt. Unless humanity gives up aggression, the robot police force that patrols the planets will destroy Earth. Klaatu and Gort reenter their spaceship, and it flies off.

SPECIAL EFFECTS

The special effects are minimal but well done. They include an excellent screen spaceship, both externally and internally, as well as the disintegration of both men and machines by the robot. In addition, a brilliant musical score utilizing electronic instrumentation was created for the film by Bernard Herrmann, and it adds much to the sinister mood of this work.

LITERARY COMMENTARY

Harry Bates' 1940 novelette "Farewell to the Master" bears only superficial resemblance to the film based upon it, as Bates himself has acknowledged. From "Farewell to the Master" the filmmakers have borrowed the spaceship, the fearsome 8-foot robot (Gnut in the story), Klaatu the benign and fully human ambassador, and two concepts: resurrecting the dead and the robot being the man's superior rather than his servant. Otherwise, the novelette and the film are quite different.

In the novelette, the spaceship "materializes" through "relativistic means," but in the film radar tracks its arrival at the relatively slow speed of 4,000 mph. Radar was invented during the interval between the novelette's publication and the film's production. The ship's appearance is the same, a smooth ovoid, and it lands in the same place, Washington, D.C. In both novelette and film, the ship is surrounded by a heavily armed military cordon and a gawking crowd. In the novelette, though, Klaatu is killed by a "madman" from the crowd rather than shot by a nervous soldier and later hunted down and killed. The robot merely freezes in position, and the "Interplanetary Wing of the Smithsonian Museum" is constructed around the robot and the ship, while Klaatu's body is interred in a mausoleum.

Much later a curious reporter, Cliff Sutherland, discovers that the robot has moved. He hides in the museum to spy and learns that it conducts experiments at night. He hears a mockingbird's song, sees the bird fly and drop to the floor dead. He hears animal sounds, sees a gorilla rush from the ship, wrestle with the robot, then drop to its knees and die. Later he hears a human voice and sees a man come from the ship. He also dies, apparently spontaneously. When Sutherland reports these events to the "Continental Bureau of Investigation," Gnut is encased in "glasstex" to immobilize him. He later melts the plastic as he does the KL 93 in the film. He continues his experimentation.

For his experiments, Gnut has created a device that can use a recording to recreate the living creature that has made the sounds recorded. He even recreates Klaatu briefly, although Klaatu tells the reporter that he, too, is dying. The theory is that each body possesses a distinctive sound; from that sound, as recreated by a recording, the body can be recreated. If the recording is flawed, the recreated body will be flawed, so Klaatu, like the mockingbird and the gorilla, is flawed and dying. Klaatu dies again, in Sutherland's arms.

The novelette contains nothing of the threat to humanity that the film possesses, and none of the strange diamonds, celestial mechanics, or stopping of electrical power. When Klaatu dies the second time, Sutherland begs the robot to tell its masters that his death was the work of a madman, that everyone on Earth regrets it. The robot then tells Sutherland that "I am the master," and the ship leaves.

Compared to the film, the novelette has a slight plot and few characters, all stereotypical. The only suspense derives from Sutherland's trying to discover the purpose of Gnut's experiments. The motivation for the ship's arrival is never clarified, nor is the motivation for its departure. In the story, Gnut has "two red eyes," but only the discovery of his midnight movements makes him at all interesting. He is bland and benevolent. In the film, Gort's visor and ray, his destruction of weapons, and his assistance to Klaatu provide him with an air of menacing omnipotence appropriate to his role of judge and potential executioner.

CHAPTER BIBLIOGRAPHY

Film References

1. Warren, B. 1982, *Keep Watching the Skies! American Science Fiction Movies of the Fifties: Vol. 1, 1950–1957* (Jefferson, NC; McFarland), pp. 19–28.

Film Reviews

1. *Christian Century* **68**, 1263 (31 October 1951).
2. *Nation* **174**, 19 (5 January 1952).
3. *New Yorker* **27**, 107 (22 September 1951).
4. *Newsweek* **38**, 90 (1 October 1951).
5. *Saturday Review of Literature* **34**, 35 (6 October 1951).
6. *Scholastic* **59**, 30 (31 October 1951).
7. *Times* **58**, 98ff (1 October 1951).

Novelette References

1. Bates, H., *Astounding Science Fiction,* 58ff (October 1940).
2. Bates, H., 1989, *They Came from Outer Space: 12 Classic Science Fiction Tales That Became Major Motion Pictures,* edited by J. Wynorski (Garden City, NY; Doubleday), pp. 93–132.

FANTASTIC VOYAGE

20th Century Fox, (US), 1966, color, 100 min.

Credits: Producer, Saul David; director, Richard Fleischer; screenplay, Harry Kleiner; based on a story by Otto Klement and Jay Lewis Bixby, as adapted by David Duncan; cinematographer, Ernest Laszlo; special effects, L. B. Abbott, Art Cruickshank, and Emil Kosa, Jr.; music, Leonard Rosenman.

Cast: Stephen Boyd (Grant), Raquel Welch (Cora Peterson), Edmund O'Brien (General Carter), Donald Pleasance (Dr. Michaels), and Arthur Kennedy (Dr. Duval).

PLOT SUMMARY

The premise of *Fantastic Voyage* is that both the US and the Soviet forces have developed the technology to "shrink an army and put it in a bottle cap," but the shrinking effect lasts only 60 minutes. One man, Benes, however, has discovered how to make the shrinking last indefinitely, and he wants to give that secret to the US. The Soviets, of course, want the secret themselves, but if they cannot get it, they want to prevent the US from getting it. Thus when Benes arrives in the US, he is attacked by Soviet agents and suffers a brain injury inoperable by normal surgical techniques. The solution is to shrink an experimental submarine, its operator, and a medical team, inject them into Benes' bloodstream, and have them perform the surgery from within his brain.

Because the surgeon who will operate on Benes from within, Dr. Duval, may be a Soviet agent, CIA agent Grant is drafted to accompany the medical team to watch Duval and prevent his killing Benes. At the huge underground CMDF (Combined Miniature Deterrent Forces) headquarters, Grant meets Captain Bill Owens, designer and operator of the small experimental navy sub; Dr. Duval, the surgeon who will use a laser to remove the bloodclot in Benes' brain; Cora Peterson, his technician; and Dr. Michaels, another physician who will act as the sub's navigator while it is inside Benes' circulatory system. This team and their vehicle will be shrunk to the size of a microbe and injected into Benes' carotid artery. They will then traverse the artery to the injury in the brain, perform the surgery, and exit through the venous system to the base of the neck, where they will be extracted. Since the sub is nuclear powered, the outside team can trace its movement by tracking the radiation source. Benes, meanwhile, has been subjected to deep hypothermia, his body temperature reduced to 28 °C and his respirations reduced to six a minute, both to reduce his risk of further damage from the blood clot and to reduce the danger to the sub and its crew.

After the sub has been injected into Benes' artery, a miniature radar array is moved into place around his head and shoulders to trace the movement of the sub. The sub at first moves quietly at 15 knots through a lovely stream of blood cells. Very soon, however, the sub becomes caught in a strong current and hurled through a fistula (a miniscule break in the artery wall) into the jugular vein. It cannot turn back because of the current and cannot go forward through the heart because of turbulence. The outside medical team decides to stop Benes' heart for 60 seconds by electrical shock to allow the sub to pass through, and then restart the heart. The sub proceeds through the right atrium of the heart and into the pulmonary artery. As it enters a capillary, an alarm signals loss of pressure in the flotation tanks caused by an electrical short. The team now has too little air to finish its journey. But being only a few cells away from an alveolar sac in the lung, they use a snorkel tube to fill the flotation tanks when Benes inhales. During the process, Grant's lifeline parts, and he is nearly lost. It could be an accident, but it could also be sabotage.

Other accidents occur, some natural and some obviously sabotage. The laser is found loose and damaged, but Grant and Dr. Duval scavenge a transistor and wire from the radio, saving the laser but leaving them without communication to the outside. As the sub enters the lymphatic system, they encounter antibodies attacking bacteria, which slows their progress. Because time is running out, they decide to take an alternate route through the inner ear. Here they discover that reticular fibers from the lymphatic system have clogged the sub's intake vents, causing overheating and power loss. Grant and the others go outside to clear the vents, but as they finish, a loud noise outside causes vibration and turbulence. Cora is swept into fibers and caught there. Grant rescues her; however, the damage she has caused brings a throng of antibodies to attack her. They cling to her suit and threaten to suffocate her until the men pull them off.

Time is now dangerously short, even though they are now near the brain. Dr. Michaels insists they do not have enough time and must return immediately to the retrieval point before they begin the growth process. Grant, however, cuts the power and tells Duval to get the laser. Grant accompanies Dr. Duval and Cora to the site of the injury, where Dr. Duval uses the laser as a rifle to burn away as much of the blood clot as possible, at least enough to free the central nerve.

Aboard the sub Dr. Michaels lures the captain out of his seat, hits him over the head, takes the controls, and heads the sub directly for the nerve. Grant seizes the laser from Dr. Duval and uses it to damage the sub. White corpuscles arrive to

Fantastic Voyage. Two crew members return to a miniaturized submarine, which is located in a blood vessel.
(Photo: Museum of Modern Art/Film Stills Archive. Courtesy of 20th Century-Fox.)

clean up the invasive material, and although Grant manages to rescue the sub captain, Dr. Michaels is pinned in the wreckage and ingested by a white corpuscle.

With only seconds left to escape Benes' body before the growth process begins, the remaining four take "the quick way out." They follow the optic nerve to the corner of the eye, where they exit just in time, are retrieved in a teardrop on a glass slide, and carried instantly to another room. Here they expand to normal size. The film ends here, with no further news of Benes' condition.

SPECIAL EFFECTS

The special effects in *Fantastic Voyage* include the shrinking of the medical team and sub, but even more importantly, the visualization of physiological details such as blood corpuscles and platelets, cell walls, heart valves, lung tissue bearing impurities, reticular fibers, the ear drum, dendritic brain tissue, and flashing synapses. All are well done.

LITERARY COMMENTARY

Isaac Asimov's 1966 novelization of *Fantastic Voyage* follows Harry Kleiner's screenplay closely in essentials such as Benes' special knowledge, his arrival by plane, his injury, Grant's joining a medical team for miniaturization and surgery, the trials within Benes' body, and the successful surgery. What Asimov adds to the basic plot of *Fantastic Voyage* is contextual detail, character development and motivation, more biological information, a heightened sense of mystery and suspicion, and a fuller and more conclusive ending.

The contextual detail added to the novel includes such items as a more futuristic limousine powered by a turbo engine and operating as a hydrofoil. This contributes to the sense of a society far enough in the future to have achieved CMDF. He also adds two laser models that Cora Peterson, as the technician, maintains, repairs, and adjusts. Dr. Duval's repairing the laser in the film is an anomaly, presumably to heighten suspicion of him, because he is a surgeon who brought a technician to handle such technical work for him.

The same buildup of detail develops more believable characters and provides motives for them. In the film, for example, Dr. Michaels is merely a pleasant, mild-mannered physician with a touch of claustrophobia and perhaps a tendency to overreact to crises. His only possible motive for sabotaging the mission is being a Soviet agent. In the novel, however, he devotes himself to medicine, to making maps of patients' circulatory systems—to the point that he always has a new map to show anyone he can get to listen. He believes firmly in helping people, and in using the miniaturization technique for medical purposes. His motive for sabotaging the mission, therefore, becomes a believable desire to prevent military use of the advanced miniaturization technique for killing people.

Cora Peterson, too, becomes more than an attractive woman who keeps remarking, "Isn't this exciting," as the sub journeys through Benes' circulatory system. In the novel she is a technician with a master's degree who works with Dr. Duval, arrogant, exacting, and slavedriving as he is, because she learns so much from him every day. She writes letters home about how much she learns working with him. This curiosity, this fascination with learning, provides her motive for volunteering for the mission and for remarking on how exciting it is.

The overall effect of the additional development in the novel is to heighten the sense of mystery and suspicion. Asimov even includes the suspicion that Benes has defected because he knows that they have the secret of indefinite miniaturization and he wants both sides to have it and thus maintain the balance of power. If he dies before he can speak, the US loses the race. Michaels is one of the suspects because of his disagreement about military use of the technique. Everyone on the sub, in fact, is a suspect except Grant.

The novel's emphasis on detail also provides greater physiological detail than the film, which relies solely on images supplemented by occasional statistics from Michaels on how many miles long the circulatory system is or how antibodies lock to their victims. In the novel Asimov explains how surface tension at the liquid-gas interface prevents insertion of the snorkel into the lungs when the team tries to refill the sub's flotation tanks. He adds detail to the description of the pleural membrane, explaining how it lies in folds touching each other but nonetheless providing room for the microscopic sub to pass between. He even discusses the difference in appearance between the blood serum and the lymphatic fluid. This additional physiological information, of course, forms part of the increased contextual detail referred to earlier, but it also makes the novel appear more fully conceptualized and finished than the film.

The additional detail at the conclusion of the novel creates the same effects. Whereas the film ends with the survivors exiting through the corner of Benes' eye in a tear and expanding to their full size, the novel includes Grant's explanation of why his suspicion fell upon Michaels, Benes' recovery, Dr. Duval's dedicated resumption of his work, and Grant's and Cora's romantic interest in each other. The film hints at a developing relationship between the two; the novel actually represents it.

CHAPTER BIBLIOGRAPHY

Film References

1. Baxter, J. 1970, *Science Fiction in the Cinema* (New York, A. S. Barnes).
2. Brosnan, J. 1978, *Future Tense: The Cinema of Science Fiction* (London, MacDonald and Jane's).
3. Sobchack, V. C. 1980, *The Limits of Infinity: The American Science Fiction Film, 1950–75* (Cranbury, NJ; A. S. Barnes).

Film Reviews

1. *Commonwealth* **84**, 557 (2 September 1966).
2. *Esquire* **65**, 112–113 (May 1966).
3. *Life* **61**, 16 (23 September 1966).
4. *National Review* **18**, 1178–1179 (15 November 1966).
5. *New Republic* **155**, 34–35 (8 October 1966).
6. *New Yorker* **42**, 225 (17 September 1966).
7. *Newsweek* **68**, 68 (29 August 1966).
8. *Saturday Review* **49**, 38 (3 September 1966).
9. *Senior Scholastic* **89**, 26 (7 October 1966).
10. *Seventeen* **25**, 116–117 (October 1966).
11. *Time* **88**, 103 (9 September 1966).

Novel References

1. Aldiss, B. W. and D. Wingrove, 1988, *Trillion Year Spree: The History of Science Fiction* (New York, Avon).
2. Asimov, I. 1966, *Fantastic Voyage* (New York, Bantam).

3. 1979, *The Science Fiction Encyclopedia,* edited by P. Nicholls (Garden City, NY; Doubleday).

Novel Reviews

1. *Booklist* **63**, 364 (15 November 1966).
2. *Books and Bookmen* **11**, 40 (September 1966).
3. *Fantasy and Science Fiction* **31**, 57 (November 1966).
4. *Horn Book* **42**, 587 (October 1966).
5. *Kirkus Reviews* **34**, 86 (15 January 1966).
6. *Library Journal* **91**, 2227 (15 April 1966).
7. *National Review* **18**, 478 (17 May 1966).
8. *New York Times Book Review* **71**, 22 (27 March 1966).
9. *The Observer,* 27 (26 June 1966).
10. *Punch* **251**, 33 (6 July 1966).
11. *Saturday Review* **49**, 33 (26 March 1966).
12. *Times Literary Supplement,* 497 (2 June 1966).

FORBIDDEN PLANET

MGM (US), 1956, color, 98 min.

Credits: Producer, Nicholas Nayfack; director, Fred M. Wilcox; story, Irving Block and Allen Adler; screenplay, Cyril Hume; cinematographer, George Folsey; special effects, A. Arnold Gillespie, Warren Newscombe, Irving G. Ries, and Joshua Meador; music, Louis and Bebe Barron.

Cast: Walter Pidgeon (Dr. Morbius), Anne Francis (Altaira), Leslie Nielsen (Commander Adams), Warren Stevens (Dr. Ostrow), Jack Kelly (Lt. Farman), Earl Holliman (the cook), and Robby the Robot.

PLOT SUMMARY

In A.D. 2200 a military rescue expedition from Earth arrives at Altair IV, a planet of the main sequence star Altair, to discover the fate of a prior scientific expedition. The only surviving expedition member is a philologist, Dr. Morbius. Dr. Morbius has constructed a remarkable robot, Robby, using the technology of the now-extinct race, the Krell, who were technologically and ethically a million years ahead of the human race. In fact, they had traveled to Earth and brought back specimens of earthly animals still present on Altair IV. The Krell race had vanished on a single night some 200,000 years earlier, when they stood on the brink of their greatest technological achievement, according to Dr. Morbius.

Dr. Morbius tells his would-be rescuers that a mysterious monster killed all the members of the prior expedition except his wife and himself. His wife later died of natural causes, after giving birth to their daughter, Altaira, nicknamed Alta. Morbius fears that the monster will return and urges his rescuers to leave.

The rescue expedition is led by Commander Adams, who, along with the ship's physician, Dr. Ostrow, is given a guided tour by Dr. Morbius of a gigantic Krell machine. The machine occupies 8,000 miles^3 and is powered by 9,200 thermonuclear reactors. This gigantic machine has maintained itself over the 2,000 centuries since the Krell vanished. Adams and Ostrow learn that Morbius has permanently increased his IQ by using a Krell machine that had been used to educate Krell children. This machine allows the user to create a replica of any object being thought of: the object is recreated each microsecond. It uses as its power source the vast Krell thermonuclear reactor system. A series of dials along the walls in a Krell laboratory measures the energy supplied by the Krell power plant. Each dial has a scale that is 10 times larger than the previous dial. As Morbius uses the Krell educational machine, only the first dial registers the consumption of energy.

The rescuers are repeatedly attacked by a monster reminiscent of the one that earlier killed the expedition members. Their nuclear weapons are useless against it. They eventually surmise that the monster is being recreated each microsecond by the great Krell machine using the thermonuclear power plants. Dr. Ostrow takes a brain "boost" from the Krell educational machine and then explains the mystery before he dies. The Krell machine converts any mental image into actual material substance, using the vast thermonuclear power plant system as its energy source. But the Krell had forgotten that they too had a subconscious, including an id. "Monsters from the id" destroyed them.

Dr. Morbius' id was the source of the monster that killed his colleagues 20 years earlier and that now threatens to destroy both the rescue expedition and his own daughter, who wishes to return to Earth with Commander Adams, whom she loves. Nothing can stop the monster. The film vividly portrays the use of vast amounts of energy as row after row of dials are lighted, showing the energy consumed by the Krell machine to create the monster. Dr. Morbius orders Robby to kill the monster, but a built-in injunction against killing any human paralyzes the robot because it knows that the monster originates in Dr. Morbius' own subconscious. At last, when Morbius realizes the truth, he renounces the creature. Dying, Dr. Morbius asks Commander Adams to throw a switch and then tells him that this has activated a chain reaction that will destroy the planet in 24 hours. The rescue party, Altaira, and Robby escape in the nick of time.

SPECIAL EFFECTS

The model work is outstanding. The spaceship and the robot sequences are truly extraordinary for 1956. There is moderate matte work, and the backgrounds, with fine studio-crafted interiors, are exceptional. The film is a forerunner of the later *2001* and the *"Star Trek"* television series.

LITERARY COMMENTARY

William Shakespeare's play *The Tempest* provided the storyline for the film *For-bidden Planet*. In the play, Prospero and his daughter, Miranda, are living on an isolated island where Prospero can study and practice his magic. Ariel, an airy spirit, is Prospero's good servant. Already on the island when they arrived was Caliban, the son of a witch and a devil, deformed in appearance and lustful, jealous, greedy, and destructive in action. For the film, the island becomes the planet Altair IV; Prospero the magician becomes Morbius the scientist; magic becomes Krell science; Robby the Robot takes the place of Ariel; and Caliban is transformed into the "monster from the id."

In the play, a storm casts a ship onto the island and introduces more characters. Among them is Ferdinand, who falls in love with Miranda. The ship also carries Stephano and Trinculo, low characters who lust after Miranda, plot to kill Prospero, and lose themselves in drunken revelry. In the film Commander Adams, as the Ferdinand character, falls in love with Altaira. Lt. Farman and the cook act the roles of Stephano and Trinculo: Farman attempts to seduce Altaira and the cook gets drunk; both of them lie easily. The "monster from the id" controls them both.

At the conclusion of the play, Prospero renounces his magic, frees Ariel, acknowledges responsibility for Caliban, and returns to the mainland with those from the ship. His "ending is despair," however, unless God's judgement reprieves him. Ferdinand and Miranda will be married, and their ending is thus a happy one. The film concludes more darkly because Morbius refuses to abandon the Krell science or to acknowledge the "monster from the id" until it is too late to save himself. He and his "magic" thus die together, although the happy ending survives in the escape of the rescue expedition from Altair IV, with the freed Robby serving as ship's navigator, and with Altaira and Adams, who will be married.

Some of the film's peculiarities probably derive from the mingling of a 17th-century play, Freudian psychology, and a futuristic interpretation of both. Magic can be either good (white magic) or bad (black magic), the two kinds distinguishable only by their respective sources (angels or demons) and their effects (beneficial or harmful). All magic is suspect because of the near impossibility of telling white magic from black magic. The magician's purity is the ultimate determiner. Morbius has idealized the Krell, elevating them almost to deities. His motives seem to him good: protecting his daughter's innocence, saving the men of the rescue expedition from the demonic force that killed his own shipmates earlier, and preserving his own ability to learn more Krell science.

These motives are not pure, however. By preserving his daughter's purity, he also restricts her life unnaturally. He selfishly retains her companionship and an unnatural degree of control over her affection and activities. While he saves the men of the rescue expedition, he also refuses them access to the Krell science and

Forbidden Planet. Dr. Morbius renounces the "monster from the id" as the lighted dials in the Krell laboratory (background) indicate the enormous power being used to create this creature.
(Photo: Museum of Modern Art/Film Stills Archive. Courtesy of Metro-Goldwyn-Mayer.)

thus retains his mental superiority. By preserving his ability to learn more Krell science, he isolates himself still further from normal human beings and becomes less human himself.

Freud theorized three parts to the personality: the superego, the ego, and the id. The superego is like a conscience, controlling and suppressing a person's nastier desires, fears, and actions. Those suppressed desires, fears, and actions, then, are not consciously recognized, and they form the id, the "monster" within each of us. According to Freudian theory, the id often surfaces in dreams and the contents of dreams represent the dark side of a person. The conscious part of the mind, the public person known to himself and to others, forms the ego.

This Freudian concept, originally an attempt to characterize the personality of an individual, can also be applied to a group. Thus, among the group on Altair IV, the superego is Robby, with his injunction against doing harm to humans, and Altaira, who interrupts her father's destructive dream, opens the house to Adams and Ostrow, and helps them destroy the "monster from the id." The ego, the conscious part of the mind, is Adams and Ostrow, normal, intelligent people who follow orders and otherwise act only for self-preservation or for the protection of those for whom they are responsible. The id, the dark underside of the mind, surfaces in Lt. Farman and the cook, but it is most apparent in Morbius, for the monster from the id is his creation. Its raging destruction represents his subconscious desire to retain possession of his daughter, to kill those who intrude into his private preserve, and to pursue his Faustian search for godlike knowledge and power. The film implies that he also destroyed the members of the earlier crew so that he could retain the Krell knowledge and power for himself.

CHAPTER BIBLIOGRAPHY

Film References

1. Clarke, F. S., and Steve Rubin, *Cinefantastique* **8**, 4–67 (Spring 1979).
2. Morsberger, R. E., *Shakespeare Quarterly* **12**, 161 (Spring 1961).
3. Rubin, S., *Cinefantastique* **4**, 5–13 (1975).
4. Sutton, M. 1984, *Omni's Screen Flights/Screen Fantasies,* edited by D. Peary (Garden City, NY; Dolphin-Doubleday), pp. 112–116.
5. Warren, B. *1982, Keep Watching the Skies! American Science Fiction Movies of the Fifties: Vol. One, 1950–1957* (Jefferson, NC; McFarland), pp. 261–274.

Film Reviews

1. *National Parent Teacher* **50**, 39 (May 1956).
2. *New Yorker* **32**, 171 (12 May 1956).
3. *Newsweek* **47**, 98 (4 June 1956).
4. *Saturday Review* **39**, 23 (7 April 1956).
5. *Scholastic* **68**, 29 (26 April 1956).
6. *Time* **67**, 112ff (9 April 1956).

Novelization

1. Stuart, W. J. 1956, *Forbidden Planet* (New York: Farrar, Straus, and Cudahy).

HANGAR 18

Goodtimes Home Video (US), 1985, 104 min.

Credits: Producer, Charles E. Sellier; director, James L. Conway; screenplay, Steven Thornley; loosely based on a story by Tom Chapman and James L. Conway; cinematographer, Paul Hipp; special effects, Harry Woolman.

Cast: Darren McGavin (Harry Forbes), Robert Vaughn (Gordon Cain), Gary Collins (Steve Bancroft), and James Hampton (Lew Price).

PLOT SUMMARY

The film opens with an onscreen message: "In spite of official denials, rumors have continued to surface about what the government has been concealing from the American public at a secret Air Force hangar. But now, with the help of a few brave eyewitnesses who have stepped forward to share their knowledge of these events, the story can finally be told."

The plot is somewhat confused, though, because the story includes a UFO and its alien pilots hidden in Hangar 18, the information that the aliens have been on Earth thousands of years ago and are our direct ancestors, and have plans to land in many other sites on Earth in the near future. The story also incorporates the destruction of a US satellite at launch, the death of a shuttle crewmember, a false story blaming the other two crewmembers for their friend's death, several murders by the CIA, an Air Force bombing of Hangar 18 in an attempt to kill the US citizens there and destroy the evidence, and White House manipulation of the CIA, NASA, Air Force, and individual citizens in order to create a favorable atmosphere for the president's reelection campaign. The story leaps back and forth among sites in space, in Houston, at NORAD headquarters, at Bannon County, Arizona, and at Washington, D.C., before ending at Hangar 18 on Wolf Air Force Base in West Texas.

The plot begins with a US shuttle launching a satellite from its cargo bay. While a crewman enters the cargo bay to resolve a minor problem, Houston control spots a blip zig-zagging across the radar screen. Then the shuttle pilot, Bancroft, sees something just above the shuttle. The satellite is launched, heads directly toward the unidentified flying object, and explodes. Debris kills the crewman in the shuttle bay.

At this point, the scene shifts to NORAD headquarters where air search units scramble for an "object coming down in Sector 12." Another shift to Bannon County, Arizona, shows a motorist watching a fireball coming down nearby. He drives to the crash site, gapes, and races off for the sheriff. Meanwhile the search unit finds the downed UFO and removes it. When the motorist and sheriff arrive, there is no sign of the craft or its having been there. The shuttle is still preparing for reentry, its crew unaware of the UFO's crash, uncertain even whether they truly saw it at all.

The NORAD general, having sent Harry Forbes of NASA to "capture" the UFO, reports to Gordon Cain in the White House that they have radar traces of a UFO, a dead shuttle crew member, and perhaps the UFO itself. The general, Cain, and the CIA chief plan to hide the UFO in Hangar 18 in absolute secrecy and blame the shuttle pilot and copilot (Bancroft and Price) for the crewman's death. Hangar 18 is the ideal spot both because of its isolation and because it is a "lunar receiving station" large enough to conceal the UFO and is equipped for study of a "space vehicle." Cain insists upon secrecy until the presidential election is over in two weeks because the president has openly scoffed at the idea of UFOs and would look ridiculous if all this came out. When Forbes arrives at Hangar 18, the UFO is inside with its lights still blinking. There is no exterior damage, and the NASA examiners think the craft made a controlled landing.

From this point, the film shifts back and forth between the study of the UFO in Hangar 18, and Bancroft's and Price's search for vindication while the CIA tries to stop them from finding evidence of the UFO's existence and to prevent their reaching Hangar 18.

When Bancroft and Price try to view the telemetry tapes of their shuttle mission, they find them altered to remove the radar traces of the UFO. They go to Crown Point, another tracking station, and view unaltered tapes that clearly show the radar blip and its strange activity. They obtain a projection of where the ship should have come down and head for Bannon County, Arizona. There they talk to the sheriff and are followed as they go to the site, where they find cut brush covering a scorched area. Bancroft picks up fused rock. As they leave, two men confront them and demand the rock. Bancroft and Price knock them down and drive away. In the ensuing car chase, the two men crash off an overpass and die when their car explodes.

Bancroft and Price then discover that their fused rock is a piece of sandstone heated to enormous temperature. They infer that the UFO was there, and leap to the conclusion that it has been moved to a nearby lunar receiving station. They rent a car, find its brakes have been tampered with, and crash. Two more men arrive and shoot at them. Bancroft and Price steal a tank truck to escape and are pursued. Price is shot dead, and the two men are also killed in the pursuit. Bancroft continues alone toward Hangar 18.

At Hangar 18, meanwhile, the investigators have entered the ship and found two bald, blue-eyed aliens dead. They are taken to a hospital for examination. The scientists explore the ship, trying controls, and Forbes finds a "specimen," who is an unconscious woman. In the ambulance, on the way to a private clinic, she sits up and screams. The investigators find that one of the controls activates a weapon that slices through an interior wall in the hangar. Another control activates a wide-screen TV showing recordings of Earth broadcasts. The researchers find symbols within the craft identical to those on a Mexican plateau visible only from the air, and to those on pyramid walls.

As they begin a translation program, the medical report on the aliens indicates that they are nearly identical to Earth humans, showing parallel evolution. The translation of alien records, however, shows that they have previously been on Earth, have captured animals (prehumans) as slaves, and have interbred with them. The physical similarities, therefore, are not a case of parallel evolution but evidence of the aliens as the "missing link" in our own evolution. Humans are the aliens' descendants. The investigators also learn that the aliens have marked several sites on Earth for landing sites in the near future.

The Air Force general, the CIA man, and the White House staff man have by now become worried about the "implications" of five men dead, decide that they are "in too deep" to go back, and arrange to have a military plane "accidentally" crash into Hangar 18 to destroy it and all the evidence within. As the film shows bombs being loaded onto a robot-controlled plane, Bancroft arrives at Hangar 18 and is taken into the ship. He is inside it when the plane crashes and Hangar 18 goes up in flames.

The film ends with a news broadcast announcing that Harry Forbes, Bancroft, and a small crew of technicians (those inside the UFO) have survived and that, as yet, there is no official statement from the White House.

SPECIAL EFFECTS

Primary special effects in the film deal with the shuttle in space, mission control in Houston, and of course, the UFO itself. They are smoothly and effectively performed.

LITERARY COMMENTARY

Many books, both fiction and nonfiction, have been written about visitations of aliens to Earth in the past. Chapter Three discusses the concept of ancient astronauts in some detail.

CHAPTER BIBLIOGRAPHY

Film Reviews

1. *Cinefantastique* **10**, 34 (3 November 1980).
2. *Monthly Film Bulletin* **47**, 235–236 (December 1980).
3. *Variety* **299** (30 July 1980).

THEM!

Warner Brothers (US), 1954, black and white, 93 min.

Credits: Producer, David Weisbart; director, Gordon Douglas; screenplay, Ted Sherdeman, based on a story by George Worthington Yates; cinematographer, Sid Hickox; special effects, Ralph Ayers; special sound effects, William Mueller and Francis J. Scheid.

Cast: James Whitmore (Sergeant Ben Petersen), James Arness (Robert Graham), Joan Weldon (Dr. Patricia Medford), and Edmund Gwenn (Dr. Medford).

PLOT SUMMARY

As the movie opens, a little girl is wandering aimlessly in the desert. She is rescued by Sergeant Ben Petersen and taken to a trailer to see if the occupants know her. Upon arrival at the trailer, they find one of its sides pulled apart and sugar cubes scattered about; the occupants are nowhere to be found. No money has been taken. The little girl is taken to a hospital, and Petersen and another patrolman stop at a country store. There they find the owner dead and the place in shambles. Money is still in the cash register. The other patrolman is left on guard while Petersen goes to the hospital. The patrolman hears a strange, rapidly pulsating, high-pitched sound, goes outside to investigate, screams, and fires his gun.

Local police authorities are outraged by the disappearance of the patrolman and call in the FBI for assistance. Special Agent Robert Graham arrives as the coroner informs the police that the store owner not only has a broken back and skull; his body contains enough formic acid to kill 20 men. The cast of an imprint found near the trailer is sent to FBI headquarters for identification.

Washington sends two scientists (Dr. Medford and his daughter, Patricia, the younger Dr. Medford) to the scene. They are experts on insects; the older Dr. Medford is a renowned myrmecologist. At the hospital they shock the little girl out of her trancelike state by having her smell formic acid. The girl screams, "Them!"

The scientists then return to the trailer site and encounter a giant ant. The sidearms of Petersen and Graham disable the ant, but it takes a submachine gun to kill it. Dr. Medford speculates that the atomic bomb exploded nine years earlier, in 1945, at White Sands, New Mexico, may have caused mutations in the desert ants to produce these giants. The authorities then search for the nest of these giant ants. When it is located, Dr. Medford advises the authorities to attack it during the day because the heat tends to drive the ants inside the nest and thus none will escape. Cyanide gas grenades are hurled into the tunnel entrance of the nest. Graham, Petersen, and Patricia Medford, wearing gas masks and armed with a flamethrower and machine gun, enter the nest itself. After surviving an attack by some ants that had been sealed off from the gas, they reach the queen ant's chamber. To their horror, they discover that two new queens had already hatched and flown away from the nest.

The federal authorities begin a nationwide hunt for the escaped queens. All stories about UFOs, flying giant insects, and other strange events are reported to Washington. The group visits a private pilot who claims to have seen UFOs shaped like ants. They are convinced that he has seen one of the queens and her male consorts. The pilot has been detained in a mental institution, and Graham ensures that he will not spread the story when he instructs a physician to keep him there until the government determines that the pilot is well! This is part of the general government cover-up designed to avoid a nationwide panic.

Finally, one of the queens turns up on board a ship at sea. It apparently entered through an open hold while the ship was docked at a Mexican port. The entire ship is infested with the giant ants, and naval gunfire sinks it.

The second queen ant has apparently landed in Los Angeles, since 40 tons of sugar have been stolen from a boxcar. Graham and Petersen travel to Los Angeles to investigate the theft. A father, dead and missing one arm, is found in his car, but his two sons have disappeared. With information from a drunk, Graham and Petersen find the model plane that the two missing boys had been flying near the entrance to a 700-mile-long sewer system.

The authorities then announce the existence of the giant ants and prepare for a major military operation to obliterate the monsters. The idea of pouring gasoline into the sewers and setting it afire is rejected because Dr. Medford cautions that the authorities must learn whether any new queens have escaped this new nest. A large number of patrols are therefore sent into the sewer system. Petersen's unit locates the missing boys in the nest, and he dies while saving them. The military then converges on the nest, and there is a furious battle at virtually point-blank range against the ants. Only after all of the worker ants are killed do the troops enter the queen's chamber. They find that all the newly hatched queens are still in the nest. These are destroyed, and humanity is saved. The picture ends with a somber question: If the first bomb explosion produced the giant ants, what did all of the subsequent explosions lead to?

SPECIAL EFFECTS

The special effects are excellent, including outstanding miniaturized photography, model work, and faultless matte work.

LITERARY COMMENTARY

Them! has no specific literary source; however, it is a dramatic presentation of familiar science fiction themes derived from Darwinian evolution (as popularly conceived, and sometimes grossly misconceived) and modified by increasing scientific knowledge and practical experience. If radiation hazards and the consequent mutations is one theme, another is the perpetual contest between human beings and insects. Both of these themes, however, fit into the larger theme of the production of dangerous effects in the biosphere, primarily by means of scientific development and technological application without regard for potential long-term consequences.

Mutation, a concept inherent in Darwinian theory, occurs in Them! as a direct result of atomic bomb tests in the New Mexico desert, producing the giant ants. In 1930, John Taine's novel The Iron Star described a mutagenic meteor transforming the local wildlife in a region in Africa into exotic shapes and sizes that were hazardous to man. Similarly, Alfred Gordon Bennett's The Demigods, published in 1939, focused specifically on the mutation of ants into a giant form, although without the intervention of nuclear tests to explain the mutation. More recently, Keith Roberts, a British novelist, showed giant wasps resulting from the radiation of a nuclear test in his 1966 novel, The Furies.

Them! Dr. Patricia Medford is attacked by a giant ant in the New Mexican desert. (Photo: Museum of Modern Art/Film Stills Archive. Courtesy of Warner Brothers.)

As practical knowledge of radiation effects increased, stories of human mutation became more common. One of the earliest novels dealing with this theme was John Taine's *Seeds of Life* (1931; reprinted 1951), in which an irradiated man becomes a superman but fails to realize the damage he has done to the genes he passes to the next generation. After the explosion of the atomic bomb in 1945, stories about human mutation because of exposure to radiation became frequent. Between 1948 and 1950, for example, Wilmar H. Shiras published a series of stories, beginning with "In Hiding," that describe the discovery of brilliant children, vastly more intelligent and creative than normal. Their parents had been exposed to radiation while working on the development of the atomic bomb. The stories were published as a novel, *Children of the Atom,* in 1953. Using the same "superchild" theme, Jerome Bixby's 1953 story, "It's a Good Life," demonstrates what terror such a child could cause a normal community if it could enforce the gratification of its every whim.

More realistically, postatomic war stories such as Lester del Rey's *The Eleventh Commandment* (1962; revised 1970) deal with the lethal effects of radiation-induced mutation. In this novel, the gene pool of the survivors is so damaged, and the mutation rate is so swift, with such disastrous results, that the postwar church encourages limitless reproduction in an attempt to produce a stable, human-

Them! Robert Graham, Dr. Patricia Medford, and Sgt. Ben Petersen enter the queen's chamber in a giant ant nest beneath the New Mexican desert.
(Photo: Museum of Modern Art/Film Stills Archive. Courtesy of Warner Brothers.)

Ellison's "A Boy and His Dog," both published in 1969, also portray postholocaust worlds of human and animal mutations, expressing concern about the defective genes caused by radiation or about the necessity of establishing an adequate gene pool among small groups of survivors. Both have been adapted into movies.

"Mutation" novels have also dealt with the hazards of pollution and of biochemical warfare. In 1966 Frank Herbert published *The Green Brain*. This novel shows humans determined to control all insect life with insecticides, creating a "green zone" clear of pests, with the surviving insects gathering in the "red zone" in South America. There the surviving insects, mutated by the insecticides, achieve a corporate intelligence. A group of these insects cling together in an imitation of the human form, sneak into the green zone, and the war is on. Alan Scott's 1971 novel, *The Anthrax Mutation*, supposes that a biological warfare laboratory has created an especially lethal form of anthrax. Through a laboratory accident, the mutated anthrax escapes to infest common bats, which then put the entire world at hazard.

Them! started a rash of imitative films during the 1950's, but it remains the best of the "radiation-induced monster mutation" films. Some of its lesser imitations are *Tarantula* (1955) and *It Came from Beneath the Sea*, also released in 1955, in which a giant octopus mutated by radiation terrorizes the California coast. *Them!* is still by far the better and more credible film.

CHAPTER BIBLIOGRAPHY

Film References

1. Warren, B. 1982, *Keep Watching The Skies! American Science Fiction Movies of the Fifties: Vol. One, 1950–1957* (Jefferson, NC; McFarland), pp. 188–195.

Film Reviews

1. *America* **91**, 367 (3 July 1954).
2. *Catholic World* **179**, 144 (May 1954).
3. *Commonwealth* **60**, 269 (18 June 1954).
4. *Farm Journal* **78**, 141 (June 1954).
5. *National Parent Teacher* **48**, 40 (June 1954).
6. *New Yorker* **30**, 61 (26 June 1954).
7. *Newsweek* **43**, 56 (7 June 1954).
8. *Saturday Review* **37**, 27 (5 June 1954).
9. *Scholastic* **64**, 29 (12 May 1954).
10. *Time* **62**, 112 (19 October 1953).
11. *Time* **64**, 79 (19 July 1954).

THE THING

RKO (US), 1951, black and white, 87 min.

Credits: Producer, Howard Hawks; director, Christian Nyby; screenplay, Charles Kederer; based on the short story "Who Goes There?" by John W. Campbell; cinematographer, Russell Harlan; special effects, Donald Stewart; music, Dimitri Tiomkin.

Cast: Kenneth Tobey (Captain Patrick Hendry), Margaret Sheridan (Nikki), Robert Cornthwaite (Dr. Carrington), Douglas Spencer (Ned "Scotty" Scott), and James Arness (The Thing).

PLOT SUMMARY

Scotty, a journalist, enters an Officers Club in Anchorage, Alaska, looking for news from the scientific party "holding a convention" at the North Pole. Immediately afterward, a message arrives from Dr. Carrington of Polar Expedition 6 announcing that they believe an airplane of unknown type has crashed in their vicinity and they need assistance. Captain Hendry flies up to investigate, accompanied by Scotty. En route they receive a radio message indicating "some kind of disturbance up here," causing a 68° difference between radio compass and magnetic compass readings.

When he arrives at Polar Expedition 6, Hendry goes immediately to see Nikki, with whom he apparently once had an affair, and who now serves as Carrington's secretary. She takes him to see Carrington for a briefing. On November 1 the seismograph registered an explosion due east, and the magnetometer registered a deviation of 12°20′ E. Such a deviation would be possible only if the disturbing force was "equivalent to 20,000 tons of steel or iron ore at about a 50-mile radius." When Hendry says it sounds like a meteor, Carrington replies that special cameras showed a "dot moving upward, then dropping."

At the impact site, the geiger counter "goes crazy." The men speculate that the engines produced enough heat to melt the ice, which then refroze over the craft. Only a tip of the airfoil protrudes from the ice. They take metal filings from the craft and find them of unknown origin. Then they attempt to melt the ice with thermite bombs. But when they explode the bombs, the ship burns under the ice. The engines explode and the geiger counter registers only residual radiation. One man finds a "hot spot" that looks like a man over 8-feet tall. They chip it out in a block of ice and take it back to the camp. A weather front closes down communications and transportation, isolating the camp from the rest of the world.

Carrington wants to remove the body from the ice block for examination, but the others more cautiously insist that it remain frozen and under watch. As the ice clears, however, the body becomes clearly visible. When the man on watch carelessly covers the block with an electric blanket, the alien revives and escapes outside. The dogs howl, and the men look outside to see the dogs tearing at a staggering figure. They rush outside and find two dead dogs and an arm the dogs had ripped off the alien.

Carrington examines the arm and declares it to be "a kind of chitinous substance with no blood, no arterial structure, no nerve endings visible, only porous, unconnected cellular tissue and a green fluidlike plant sap, probably with a sugar base." Carrington concludes that the "Thing's" development was not handicapped by emotional or sexual factors and, therefore, it has a superior brain. He mentions "thinking vegetables" on Earth such as the telegraph vine or the acanthus century plant that catches mice, bats, and squirrels using a sweet syrup as bait. From under the soft tissue in the palm of the hand Carrington removes a seed pod. Later the arm shows a temperature rise of 20° and ingests the canine blood on it.

Carrington insists that the Thing is "our superior in every way," and he secretly grows seeds from the pod, nourishing them with blood plasma. When the Thing kills a dog, removing its blood, and then is discovered to have broken into the greenhouse, Carrington keeps the knowledge from Hendry, and places two men on watch in the greenhouse. The two men are later seen hanging upside down from the rafters with their throats cut. The military then traps the Thing in the greenhouse. Nikki reveals to Hendry that Carrington is growing seeds from the pod in his lab that reproduce with "amazing speed."

Apparently the Thing needs human blood for the nourishment of seeds that it must be growing in the greenhouse, trying to reproduce itself. The men now speculate that the Thing is trying to grow a horrible army, turning humans into food for it and its progeny.

Hendry and his men now begin serious attempts to destroy the Thing. As they contemplate how to kill a vegetable, Nikki suggests, "Boil it." The Thing then escapes from the greenhouse and attacks the group. They throw kerosene on it, and ignite the fuel. The Thing flees outside into the storm but survives. Then one of the men suggests using electricity to kill the creature. They rig a generator and cables to wire the corridor, creating an "electric flytrap." As the Thing appears in the corridor, Carrington dashes toward it to plead for friendship with this being which he believes is "wiser than anything on Earth." The Thing kills him, advances, and is electrocuted. The men burn the Thing to ashes, along with its arm and the seedlings. Hendry then gives Scotty permission to tell the story, and the film ends with the warning: "Watch the skies! Everywhere! Keep looking!"

SPECIAL EFFECTS

Special effects in *The Thing* are minimal. James Arness wears not very effective makeup that works only because he appears as the Thing in dimly lit scenes at a distance. Otherwise, the pulsing seedlings and the arctic setting are the major— and effective—special effects.

LITERARY COMMENTARY

The film of *The Thing* differs significantly from John W. Campbell's story "Who Goes There?" (1938). The polar setting, the discovery and accidental destruction of the ship, and the discovery, retrieval, and revival of the alien are similar in both film and story, but the story provides a very different and more dangerous alien, a greater terror, and a subtler means of detecting the alien's presence.

In "Who Goes There?" the setting is Antarctica instead of the North Pole; the ship has been buried in the ice for millions of years instead of just having crashed; and the thermite bombs burn the ship because of magnesium in its hull alloy instead of because its engines burned. The alien in its block of ice is not humanoid, as in the film, but a creature with three eyes, rubbery blue hair, and razor-sharp tentacles. Nor does it signal its presence by radiation. In fact, the alien *becomes* its prey, taking on the shape and character of its victim.

The Thing. Scientists study plants growing from a pod taken from the arm of an alien creature.
(Photo: Museum of Modern Art/Film Stills Archive. Courtesy of RKO.)

This shape-changing ability appears when the alien tackles the dogs. It can digest and imitate any living creature, retaining enough body mass to digest and imitate another. It reproduces with incredible efficiency merely by feeding. The problem the humans face is that one or more of them, seemingly normal and natural, has become the alien. They must devise a way to tell which of them remains human and which has become alien. With no visible, tangible, or behavioral differences apparent between human and alien, the biologist uses a blood-serum test to determine "who goes there."

He injects blood from two of the humans into one of the dogs so that its blood will become "human-immune." Later, when some of the dog's blood is drawn, separated, and human blood added to the serum, a reaction occurs to prove the added blood was indeed human. Addition of blood from any other protein source gives no reaction. This test seems infallible—until the men discover that the original samples injected into the dog were from one human and one alien. The test is worthless. And which of the two is the alien?

At this point the men become extremely paranoid, and one of them, Blair, insists that he is not staying with aliens, that he will go mad. The others isolate him in a shack designed for cosmic ray observation and continue to work on their problem: how to identify the aliens among them. They watch each other, stay in a group and try to devise an effective test.

The new test is a simple one, based on the concept that the alien is selfish and "every part is a whole," i.e., each alien blood cell is "a newly formed individual in its own right." When a human blood sample is threatened by a hot wire, the blood simply burns; but when an alien blood sample is threatened by a hot wire, the blood cells draw away from the source of danger. As each man is tested, he is accepted as human—or torn apart and his cells destroyed by electrical charges and caustic acid.

Suddenly the relieved survivors remember Blair, isolated in the shack. As they approach the shack, they see a "fiercely blue light" coming from the cracks in the shack's door, hear a soft humming and the "clink and clank of tools." Blair is a Thing. They destroy him and discover that he has, in the few days he has been in the shack, created a small atomic power source to create the blue light and the heat natural to "a hotter planet that circled a brighter, bluer sun." He had also nearly finished a knapsack-sized antigravity device with which he could have escaped Antarctica and reached the heavily populated areas of Earth. He has been destroyed just in time.

CHAPTER BIBLIOGRAPHY

Film References

1. Baxter, J. 1970, *Science Fiction in the Cinema* (New York, A. S. Barnes).
2. Brosnan, J. 1978, *Future Tense: The Cinema of Science Fiction* (London, MacDonald and Jane's).
3. Lucanio, P. 1987, *Them or Us: Archetypal Interpretations of the Fifties Alien Invasion Films* (Bloomington, Indiana; University Press).
4. 1979, *The Science Fiction Encyclopedia*, edited by P. Nicholls (Garden City, NY; Doubleday).
5. Sobchack, V. 1980, *The Limits of Infinity: The American Science Fiction Film* (Cranbury, NJ; A. S. Barnes).
6. Warren, B. 1982, *Keep Watching the Skies!* Vol. 1 (Jefferson, NC; McFarland).

Film Reviews

1. *Christian Century* **68**, 807 (4 July 1951).
2. *Commonwealth* **54**, 143 (18 May 1951).
3. *Library Journal* **76**, 883 (15 May 1951).
4. *Nation* **172**, 497 (26 May 1951).
5. *Nation* **174**, 19 (5 January 1952).
6. *New Republic* **124**, 23 (21 May 1951).
7. *New Yorker* **27**, 78 (12 May 1951).

8. *Saturday Review of Literature* **34**, 28 (21 April 1951).

9. *Time* **57**, 110 (14 May 1951).

Story References

1. 1980, *They Came from Outer Space: 12 Classic Science Fiction Tales That Became Major Motion Pictures,* edited by J. Wynorski (Garden City, NY; Doubleday).

THE TIME MACHINE

MGM (US), 1960, color, 103 min.

Credits: Producer and director, George Pal; screenplay, David Duncan, based on the novel *The Time Machine* by H. G. Wells; cinematographer, Paul C. Vogel; special effects, Gene Warren and Wah Chang.

Cast: Rod Taylor (George, the time traveler), Alan Young (David Filby), Yvette Mimieux (Weena), and Sebastian Cabot (Dr. Hilliard).

PLOT SUMMARY

The Time Machine presents a look at humanity's future dominated by recurring wars and their ultimate effect in the year 802,701. Then the time traveler finds humanity divided into the Eloi, a childlike, ignorant, and slothful people who are actually raised as food by the brutish, industrialized, but troglodytic Morlocks.

George, an inventor, holds a dinner party New Year's Eve, 1899, for several male friends, including David Filby and Dr. Hilliard. After dinner he presents a box in which is a working model of a time machine, "the results of two years' labor." He and Dr. Hilliard explain time as the "fourth dimension," which cannot be seen or felt. Then George demonstrates movement through time by activating the model of his time machine. It disappears. Dr. Hilliard claims that the future is irrevocable and cannot be changed; George says, "I wonder. Can man control his destiny? Can he change the shape of things to come?"

David Filby remains behind when the other guests leave and asks George why he is preoccupied with time. George expresses his revulsion at people's finding new ways to kill each other and says he prefers the future, implying that the future will hold peace. When David leaves, George writes a note on his calendar inviting the same group of men to dinner on 5 January 1900, the following Friday. Then he goes into his laboratory, climbs into his time machine, switches it on, and goes

The Time Machine. The time traveler prepares to move forward into the future by pushing
the lever on his time machine forward.
(Photo: Museum of Modern Art/Films Stills Archive. Courtesy of Metro-Goldwyn-Mayer.)

forward in time. At first he notices no change, then realizes that the candle is
shorter by inches and the clock reads 8:09 instead of 6:31. He continues forward.

He stops in 1917 and discovers his house abandoned, full of cobwebs, dust,
and rats, with the door boarded up and the yard full of weeds. Outside, he meets
James Filby, the son of his friend, who was killed in the war a year ago. George
continues forward to 1940, where he finds a new war fought with airplanes and
bombs. His house is gone. Moving forward again to 1966, George finds another
war. He sees old Filby in a warden's uniform, and Filby recognizes him. Then the
last siren sounds, a bomb hits, causing a volcanic explosion and magma flowing
through London streets. George flees into the future through centuries of darkness.
Suddenly the rock that had engulfed him disappears and George stops his machine
in 802,701.

In this far future George finds a monolithic building topped by a sphinx head,
but his echoing knocks on the doors bring no response. He sees flowers everywhere, trees, and vines fruit laden, and visualizes "nature tamed." Another building contains tables set for meals but is empty. Outside he hears voices, sees a
laughing, playing crowd of people. They merely watch as a woman caught in the

river screams. George rescues her. She is Weena of the Eloi. He asks her to take him to someone older, and she says there is no one older. The Eloi ignore George, but under questioning reveal that they have no government, no laws, no work, and food that "just grows." When George asks one for books, he is led to a dusty room where moldy books crumble in his hands.

George storms off to return to his machine but finds it gone. Tracks show it has been moved into the Sphinx building. This is where the Morlocks live. Indeed, shadowy figures appear, capture Weena, and George saves her again, keeping the Morlocks away with fire.

In the morning George hears the sound of machines from a well in the Earth. Weena identifies this sound with Morlocks. The "rings" told her of the machines, she says. The rings are records of the past. One tells of war between East and West, air pollution, and the failure of the last surviving oxygen factory. Another tells of individual choices: living in caverns or living in sunlight.

George determines to enter one of the wells to retrieve his time machine, but as he does a siren sounds and the older Eloi, including Weena, respond by marching to the Sphinx doors and entering. George follows, but the doors close in his face. The Eloi outside tell him the "all clear" has sounded and the doors will not open again until the siren sounds next. He returns to the vent and climbs down. In the caverns he finds moss from which he makes a torch, sees strange machines, and sees bones—the Eloi, bred like cattle for Morlock consumption. A Morlock with a whip drives the Eloi, and George grabs Weena, seizes the whip, and fights. He sets fire to the moss, and the Morlocks back off. George and the Eloi climb out the well to the surface, where they throw dead wood down to feed the fire. Explosions occur and smoke and flames burst from the wells.

The doors of the Sphinx open long enough for George to enter and find his machine. He fights off Morlocks, returns to 5 January 1900, stops in the garden, and joins the dinner party of his friends. They do not believe his story, even though he has brought a flower that is different from any species known today.

The films ends with David Filby finding George and his machine gone. David assumes he has returned to the future to find Weena and to help the Eloi build a new world.

SPECIAL EFFECTS

The special effects are unspectacular but effective. The major device is time-lapse photography to indicate the incredibly swift passage of time as George travels into the future. However, the buildings and caverns of the Eloi and the Morlocks—and the crumbling books—are also effective special effects.

LITERARY COMMENTARY

H. G. Wells' novel *The Time Machine* provided the story for the film. Although the film follows the novel reasonably closely, significant differences do exist. These differences lie primarily in the character of the dinner party, George's motive for time travel, the journey itself, the nature of the Eloi and the Morlocks, and the social cause of this split in humanity.

In the film the original dinner party is of a special group of friends to whom George demonstrates his time machine and whom he invites to dinner again five days later in order to prove that he has a working time machine. In the novel, however, George is merely a sociable fellow who regularly hosts a dinner party every Thursday. He demonstrates his model to one group of guests and returns a week later to tell his story of the future to another group of guests. There has been no challenge and proof, but merely a week's journey between the regularly scheduled dinner parties.

George's motive in the film also has been changed for dramatic purposes. In the film he wishes to travel into the future because he is horrified by war and seeks the peace and prosperity he imagines in the future. In the novel he is simply a clever inventor with a wonderful device he wants to test in a practical way by traveling into the future.

Because of the film's antiwar motif, George's journey into the future contains stops in 1917, 1940, and 1966 when he can see the escalating horrors of war. In the novel George goes straight to 802,701 without any stops along the way. He is not concerned with the horrors of war but with the test of his machine and a yearning to see what the far future is like.

The Eloi George meets in the film future are almost zombielike in the way they ignore each other and George. They are not merely unintelligent but uncurious. In the novel, by contrast, they are truly childlike. They are incessantly curious, following George about, asking questions, running and playing—but with an attention span of only minutes at the most. In addition, the Eloi are only about 4-feet tall. This makes the relationship between George and Weena more that of a father and child than the film's romantic couple.

The Morlocks of the novel also differ from those of the film. In the film they are hairy, animal-like, growling and yowling brutes. In the novel, they are small, pale, hairless, timid, and industrious. George sees them as antlike. This, in turn, means that no such battle between Morlocks and George occurs in the novel as occurs in the film despite the Morlocks' cannibalism.

Indeed, the major difference between film and novel occurs in the explanation of this peculiar future. The film would have us believe that incessant wars drove humanity underground into bomb shelters, that some humans stayed there to develop into the brutish Morlocks, and that some returned to the surface to find a kind of garden paradise where they flourished except for predation by the Morlocks. The Morlocks thus could use the siren to lure the Eloi to "shelter" in their caverns because of the racial memory of war, and they could sound the all clear to stop the Eloi from entering the caverns.

This explanation almost directly opposes the explanation given in the novel. There Wells postulates a future of peace and prosperity in which the aristocratic class, the "idle rich," become even more idle and rich until Darwinian selection reduces them to the childlike Eloi, idling their time away in the garden. The

Morlocks, busy little possessors of underground industrial complexes, represent the natural development of the working class. Kept in dark "sweatshops" further and further from the homes and gardens of the rich, working all their lives, they developed through natural selection into the pallid, ever-busy, cave-dwelling Morlocks. Whereas the film presents recurring war as the cause of the Eloi passivity and the Morlock brutishness, the novel presents the absence of conflict as the cause.

CHAPTER BIBLIOGRAPHY

Film References

1. Baxter, J. 1970, *Science Fiction in the Cinema* (New York, A. S. Barnes).
2. Brosnan, J. 1978, *Future Tense: The Cinema of Science Fiction* (London, MacDonald and Jane's).
3. Meyers, R. 1980, *The World of Fantasy Films* (New York, A. S. Barnes).
4. 1979, *The Science Fiction Encyclopedia*, edited by P. Nicholls (Garden City, NY; Doubleday).
5. Sobchack, V. 1980, *The Limits of Infinity: The American Science Fiction Film* (Cranbury, NJ; A. S. Barnes).
6. Warren, B. 1986, *Keep Watching the Skies: American Science Fiction Movies of the Fifties: Vol. II: 1958–1962* (Jefferson, NC; McFarland).

Novel References

1. Hillegas, M. R. 1967, *The Future as Nightmare: H. G. Wells and the Anti-Utopians* (New York, Oxford University Press).
2. Huntington, J. 1982, *The Logic of Fantasy. H. G. Wells and Science Fiction* (New York, Columbia University Press).
3. Murray, B. 1990, *H. G. Wells* (New York, Continuum).
4. 1972, *H. G. Wells: The Critical Heritage*, edited by P. Parrinder (London, Routledge and Kegan Paul).
5. Wells, H. G., *The Time Machine* (1895), *Seven Science Fiction Novels of H. G. Wells* (New York; Dover).

TOTAL RECALL

Carolco Pictures (U.S.), 1990, Color, 113 min.

Credits: Producer, Buzz Feitshaus and Ronald Shusett; director, Paul Verhoeven; screenplay, Ronald Shusett, Dan O'Bannon, and Gary Goldman, based on "We Can Remember It for You Wholesale" by Philip K. Dick; cinematographer, Jost Vacano.

Cast: Arnold Schwarzenegger (Douglas Quade), Sharon Stone (Laurie Quade), Ronny Cox (Kohagen), Rachel Ticotin (Melina), and Michael Ironside (Richter).

PLOT SUMMARY

The film opens with a Marscape in which two figures in pressure suits hold hands. The man falls, breaks his faceplate, and decompresses. At that point, Douglas Quade, a happily married laborer on Earth, wakes up screaming. He and his wife, Laurie, are in bed, and he has had a nightmare. The next morning Quade watches a news report about terrorists on Mars who have demanded independence, temporarily halted extraction of the "turbinium" necessary for Earth's weapons, and used bombs in an attempt to reopen the sealed Pyramid Mine. He says, "Laurie, let's move to Mars," but she discourages him.

On the way to work Quade sees an ad for Recall, Inc., which promises to implant memories of a vacation to Mars as real as if he had been there. He chooses the "secret agent" identity and a woman who is "sleazy and demure." When the doctor tries to implant this memory, however, Quade has a "schizoid embolism" and becomes violent. The Recall personnel realize that someone has erased his memory, and they erase any memory of his visit to Recall, then "dump" him outside.

As he returns home, Quade encounters a coworker who asks about his "trip to Mars." Then a group of men try to kidnap Quade. He kills them all and returns to Laurie, who claims he is having paranoid delusions until he shows her the blood on his hands. Then she tries to kill him. She claims that their marriage is an implanted memory, that she never saw him before six weeks ago. "Your whole life is just a dream," she says. When a TV monitor shows more men coming through the lobby, he slugs her and runs. The men use a "bug" in his skull to trace him. Their leader is Richter, who takes orders from Vilos Kohagen, Mars Administrator and its virtual dictator. Richter loves Laurie.

After a violent pursuit through the subway terminal and on a subway train, during which several innocent bystanders are killed, Quade checks into a hotel. Here a strange man calls him on the videophone and tells him that he is bugged, that he must wet a towel and wrap it around his head to muffle the signal. The wet towel works and Richter loses the signal. The stranger then asks him to look out the window, shows him a metal suitcase that had been left with the stranger for safekeeping. The suitcase contains money, ID cards, a holographic projector, and a communicator with a message from "Hauser," his former identity. Hauser says he worked for Mars Intelligence. He instructs Quade to remove the bug from his nostril so that his movements cannot be traced. Hauser tells Quade to go to Mars and flash the Brubaker ID at the desk in the Hilton Hotel.

The rest of the movie takes place on Mars. A large woman being questioned by a customs officer is Quade wearing a mask, which malfunctions. He removes his mask and tosses it toward the police. After it explodes, Richter shoots at Quade and the bullets shatter the dome, decompressing the area and blowing people into the thin Martian atmosphere. Quade escapes to reach the Hilton Hotel.

Richter meets with Kohagen who tells him that Kuato, the rebel leader, seeks the information inside Quade's head and may retrieve it because Kuato is psychic. Quade, meanwhile, finds a note left by Hauser telling him to go to the Last Resort in Venusville and ask for Melina. A hustling cabby takes him there, explaining as another explosion occurs, "That's the rebels." The cabby also tells Quade as they pass misshapen people on the street that these are mutants, that the mutants are psychics, the result of cheap domes, bad air, and no way to screen out radiation. When they reach Melina, she recognizes Quade as Hauser, does not trust him, and tosses him out.

Back at the hotel, a "Recall doctor" enters Quade's hotel room to try to convince Quade that he is really at Recall on Earth and suffering a "free-form delusion" based on their memory tapes. Laurie then enters the room to reinforce this idea. Quade realizes that they are both deceiving him and shoots the doctor, but more men appear and capture him. Laurie kicks him around, but then Melina appears, shoots Quade's captors, and fights with Laurie. Quade shoots Laurie as she is about to kill Melina. Melina leads him—and the cab driver—to Kuato, still pursued by Richter and his men. Kohagen orders "Sector G" sealed and cuts the air inflow to suffocate his opposition.

In the catacombs Quade meets Kuato, who tells him a man is defined by his actions, not his memory. Quade then remembers the alien machine in the Pyramid Mine. Richter and his men break in, the cabby reveals himself as Kohagen's agent, and Quade and Melina are captured. Kohagen claims that Hauser planned Quade as the perfect mole to destroy the rebellion and hooks Quade and Melina into a psychprobe console to give then new identities. Quade breaks loose, frees Melina, and heads for the Pyramid Mine. The alien artifact, located in the mine, is an air generator that will melt the ice core of Mars and thus supposedly release oxygen to provide an atmosphere for the entire planet.

After more mayhem, during which Quade and Melina use the holographic projector to eliminate a small army trying to kill them, Kohagen activates a bomb which depressurizes the mine. Kohagen is blown out to decompress in the rarefied Martian atmosphere. After Quade starts the reactor, he and Melina, too, are blown out and decompress. However, the miracle machine produces a breathable atmosphere so quickly that Quade and Melina recover entirely, and the movie ends with them kissing under the blue skies of Mars.

SPECIAL EFFECTS

The main special effect are the mask Quade wears when entering Mars, the holographic projector, and the "terra-forming" of Mars.

LITERARY COMMENTARY

The film *Total Recall* derives loosely from Philip K. Dick's short story "We Can Remember It for You Wholesale." The film and the story differ significantly, beginning with the main character's name ("Quade" in the film and "Quail" in the story) and occupation (construction worker in the film and "clerk" in the story), and then diverging wildly. The film emphasizes action, violence, government conspiracy, victimization and mutation of the masses on Mars, and their salvation through the heroic actions of Quade fighting against the entire military might of Mars. The story, on the other hand, contains no violence or conspiracy, no Martian victims, or heroic salvation. The emphasis in the story lies with the discovery and significance of buried memories.

In "We Can Remember It for You Wholesale" Quail has the same fascination with Mars that Quade does in the film, and like Quade, goes to Recall, Inc., for an implanted memory of two weeks on Mars with the "secret agent" option. As in the film, Recall, Inc., discovers that Quail really had been a secret agent on Mars, and they attempt to bury the memory and turn him loose. The memories resurface, however, and Quail becomes hunted by government agents.

At this point film and story part company. Instead of a violent pursuit as in the film, Quail never returns home, never picks up a gun, and never kills anyone. His "implant" is a telepathic one that cannot be removed, and through it he strikes a deal with his pursuers. They will try to bury the unwanted memory with another session at Recall, Inc., to give him an even stronger memory to override the unwanted one. At Recall, Inc., a psychologist searches for his strongest fantasy upon which to build the new memory. He discovers that when Quail was eight he fantasized saving Earth from an invasion by small, utterly vicious, ratlike aliens. He did so by treating them so humanely and kindly that they were awed by his gentleness and agreed to leave Earth alone as long as he lived. Quail's fantasy of being the unknown, unsung hero, secret savior of Earth because of this gentleness and humaneness becomes the memory that Recall, Inc., will attempt to implant. However, as the doctors put Quail under "narkidrine" to implant this memory, they discover that the "fantasy" is reality. Quail did meet such aliens, impress them with kindness, and thus save Earth. As long as he lives, Earth is safe; if he is killed, Earth will be invaded.

Although the film and the story both use the device of buried memories recovered and altering the perceived identity of the main character, the film celebrates violence and machismo; the story celebrates gentleness and humanitarian values.

CHAPTER BIBLIOGRAPHY

Film References

1. Anthony, 1989, *Novelization of the Screenplay* (New York, Avon).
2. Griffin, N., *Premiere* 70–74 (June 1990).
3. Meisel, M., *Film Journal* **93**, 12–13 (April–May 1990).
4. Schwarzenegger, F. B., *Cinema Fantastique* **20**, 8–9 (5 November 1990).
5. Seldenberg, R., *American Film* **15**, 34–39 (June 1990).

Film Reviews

1. *Ad Astra* **2**, 48 (September 1990).
2. *American Film* **15**, 38–39 (June 1990).
3. *Commonwealth* **117**, 456 (10 August 1990).
4. *Film Comment* **26**, 24–29 (July–August 1990).
5. *Gentlemen's Quarterly* **60**, 79–80 (June 1990).
6. *Maclean's* **103**, 67 (11 June 1990).
7. *New York* **23**, 68 (18 June 1990).
8. *The New Yorker* **66**, 92 (18 June 1990).
9. *Newsweek* **115**, 62 (11 June 1990).
10. *People Weekly* **33**, 11–12 (18 June 1990).
11. *Rolling Stone*, 42 (12–26 July 1990).
12. *Time* **135**, 85 (11 June 1990).
13. *Time* **136**, 64 (2 July 1990).
14. *Variety* **339**, 24–25 (6 June 1990).
15. *Video* **14**, 70 (December 1990).
16. *Village Voice* **35**, 69–70 (12 June 1990).

Story References

1. Aldiss, B. W., and D. Wingrove, 1988, *Trillion Year Spree: The History of Science Fiction* (New York, Avon).
2. Dick, P. K., *Fantasy & Science Fiction* (April 1966); reprinted in 1987, *The Little Black Box. The Collected Stories of Philip K. Dick* Vol. 5 (Los Angeles, Underwood/Miller), pp. 157–175.
3. Greenberg, M. H., and J. D. Olander 1983, *Philip K. Dick* (New York, Taplinger).
4. Mackey, D. A. 1988, *Philip K. Dick*, Twayne's United States Authors Series (Boston, Twayne), p. 533.
5. 1979, *The Science Fiction Encyclopedia*, edited by P. Nicholls (Garden City, NY; Doubleday).
6. Warrick, P. S. 1987, *Mind in Motion: The Fiction of Philip K. Dick* (Carbondale, Southern Illinois University Press).

Films Without Literary Commentary

THE ABYSS

Twentieth-Century Fox (US), 1989, color, 140 min.

Credits: Producer, Gale Anne Hurd; director, James Cameron; story, James Cameron; cinematographer, Mikael Salomon; music, Alan Silvestri; supervision of visual effects by John Burno, Hoyt Yeatman, Dennis Muren, Robert Skotak, and Gene Warren Jr; special visual effects, Dream Quest Images, Industrial Light and Magic, and Fantasy II Film Effects.

Cast: Ed Harris (Bud Brigman), Mary Elizabeth Mastrantonio (Lindsey Brigman), and Michael Biehn (Lt. Coffey).

PLOT SUMMARY

The film opens on board a US nuclear submarine as the crew is trying to track a mysterious underwater object that moves incredibly fast. The object seems to be headed directly for them, but at the last minute avoids a collision. However, due to the object's wake and an unexplained power failure, the submarine hits an underwater cliff, severely damaging the vessel and killing the crew. The Navy requests that the crew of a nearby experimental deep-sea oil rig assist in exploring the wreckage. Three Navy Seals are assigned to go on board the submersible rig, accompanied by the rig's designer, Lindsey Brigman, a hard-nosed scientist who is married to, but estranged from, Bud Brigman, the rig crew's supervisor. As the four descend 1,700 feet to the rig, Lindsey warns the Seals to look for signs that the submersion has affected their nervous systems. The Navy Seals' crew leader, Lt. Coffey, displays the symptoms that Lindsey described but hides them from the others.

Later, the Seals and the oil rig crew enter the submarine's wreckage, finding many dead sailors. One of the crew sees something unexplainable and panics, causing him to accidentally maladjust his oxygen gauge and thereby go into a coma. Meanwhile, Lindsey sees a strange flash of light that she cannot explain.

Back on board the rig, the crew learns that a powerful hurricane is headed toward them and that they must disconnect their rig from the crane on the platform that floats on the ocean's surface. However, just as the crew attempts to board their underwater vessel in order to detach the cable from the crane, the vessel suddenly submerges. The Navy Seals have requisitioned it to return to the submarine's wreckage: they soon return with a mysterious object.

Because of the delay that is caused by the Seals' use of the underwater vessel, the rig crew does not have enough time to detach the cable. The crane is pulled

The Abyss. Bud Brigman prepares to descend into the abyss to deactivate a nuclear bomb.
(Photo: Museum of Modern Art/Film Stills Archive. Courtesy of 20th Century-Fox.)

off the platform and sinks into the water, landing near the rig. The crane then topples into the underwater abyss located next to the rig, which is damaged by being dragged by the crane.

When Lindsey goes outside the rig to assess the damage, she sees unexplainable objects that resemble fluorescent stingrays. She photographs the objects, but the crew dismisses her report as imagination.

The crew discovers that the mysterious object brought by the Seals from the sunken submarine is a nuclear warhead. It becomes obvious that Lt. Coffey is starting to go insane. The crew is then visited by an alien water creature, and all but Coffey are now convinced that these creatures exist and are friendly. Coffey, thinking the aliens may be some kind of Soviet invention designed to steal the nuclear warhead, decides to destroy the creatures. He locks up the crew and prepares to descend in one of the underwater vessels to drop the nuclear warhead into the abyss from which the creatures originate. The crew free themselves, and Bud Brigman and another crew member swim under the rig to the underwater vessel's dock and fight Coffey, but are unable to prevent him from escaping with the missile. Bud and Lindsey Brigman then pursue him in another vessel. As the two vessels ram one another, Coffey and the warhead both fall into the abyss.

The submersible piloted by Bud and Lindsey Brigman is damaged and unable to return to the rig. They discover that they have only one diving suit on board the submersible. Both cannot make it back to the rig. Lindsey's solution to their crisis is to have Bud don the diving suit and carry her back to the rig, hoping that the ice-cold water will prevent irreversible damage to her brain caused by her drowning. At first the crew is unable to revive Lindsey, but Bud refuses to accept that his wife is dead and finally revives her.

Bud then volunteers to descend into the abyss using the Navy's experimental liquid oxygen tank. He drops to 17,000 feet and deactivates the warhead, but is unable to return to the rig on the remaining liquid oxygen. As he lays dying, an alien creature approaches and leads him to an Atlantis-type city. There he is placed in a room where he recovers from his ordeal while the aliens begin to communicate with him.

Meanwhile, the rig is contacted by the Coast Guard on the surface, and they try to work out a plan to rescue the crew before their oxygen supply is depleted. The rig crew suddenly receives a message from Bud, who informs them that the alien underwater city is being moved to the ocean's surface. The rig is lifted atop this underwater city and the sudden ascent of 1,700 feet by the rig mysteriously does not result in killing its occupants.

THE ADVENTURES OF BUCKAROO BANZAI

Twentieth-Century Fox (US), 1984, color, 103 min.

Credits: Producers, Neil Canton and W. D. Richter; director, W. D. Richter; screenplay, Earl MacRauch; cinematographer, Fred Koenekamp; special effects, Michael Fink; music, Michael Boddicker.

Cast: Peter Weller (Buckaroo Banzai), Jeff Goldblum (New Jersey), Lewis Smith (Perfect Tommy), Ellen Barkin (Penny Priddy), Robert Ito (Professor Hikeda), and John Lithgow (Dr. Emilio Lizardo/Lord John Whorfin).

PLOT SUMMARY

The picture commences with the research team of nuclear physicist Dr. Buckaroo Banzai readying his jet car for what is to be a speed test. Meanwhile, Banzai, who is also a neurosurgeon and a rock star, is performing a delicate brain operation with the assistance of another doctor, New Jersey, whom Banzai invites to join his rock band.

When the surgery is completed, Banzai enters the jet car, which he accelerates to 700 mph. He then conducts the second part of his experiment, which is a surprise even to the government agency sponsoring his research—namely, to drive the jet car through a solid mountain. This feat is accomplished with the assistance of an oscillation overthruster, a device which weakens the bonds between the nuclei and their electrons so that the car can pass through seemingly solid rock.

During its passage through the mountain, the jet car is struck by something. After he emerges from the mountain, Banzai finds living tissue under the jet car.

We next see Banzai performing at a nightclub with his rock group, the Hong Kong Cavaliers. One of the customers, Penny Priddy, is very depressed and tries to shoot herself. Banzai soon discovers that Priddy, an orphan, is the twin sister of his dead wife. He persuades the police to release her into his custody.

The Banzai team next holds a news conference at which they describe the jet car's passage through seemingly solid matter: Banzai notes that the protons, neutrons, and electrons in matter occupy only one-quadrillionth the volume of matter. He asserts that the living tissue which struck his car came from another dimension. We also learn that the oscillation overthruster is the result of years of work by his colleague Dr. Hikeda, and Dr. Emilio Lizardo.

Dr. Lizardo, whose earlier experiments had resulted in a brief penetration into the eighth dimension during which his body was taken over by Lord John Whorfin, escapes from a mental institution upon hearing the news of Banzai's passage through the mountain.

During the conference, Banzai is interrupted by a call, ostensibly from the President, which actually originates from an orbiting alien spaceship. When Banzai answers the telephone, he receives an electrical shock that alters his body chemistry so that he can see that there are alien beings in the audience who are seen as normal humans by everyone else. These aliens then abduct Prof. Hikeda. Banzai follows them on a motorcycle and rescues Hikeda.

Meanwhile, the alien spaceship sends a probe to Earth in order to give Banzai further information. The probe crashes and two of its three crewmembers are killed, but the third reaches Banzai's headquarters. We learn that the ship is from the eighth-dimension world of Planet 10. The former dictator of that world, Lord Whorfin, and a handful of his followers (called Red Lectroids) are now threatening to return using the oscillation overthruster. In order to prevent that from happening the aliens (called Black Lectroids) in the spaceship threaten to initiate a world war between the US and the USSR in order to destroy Whorfin.

The Red Lectroids attack the Banzai compound and capture Penny Priddy, who has the overthruster. Banzai and his friends learn that Whorfin's followers came to Earth in October 1938, landing in New Jersey, and that they are the inspiration for "The War of the Worlds" broadcast by Orson Welles that Halloween eve. The aliens operate a company, Yoyodyne, in which they have built a spaceship to return them to their own dimension. They need the overthruster to permit the ship to return to Planet 10.

Banzai's forces storm Yoyodyne, but the Red Lectroids' ship takes off, without Banzai's overthruster. Lizardo/Whorfin intends to return to Planet 10 using an overthruster of his own design, but his ship is destroyed in aerial combat with a ship piloted by Banzai. In gratitude for Banzai's efforts, the friendly Black Lectroids revive Penny Priddy, who had been killed by the Red Lectroids. Thus, the movie ends on a happy note and we are promised further adventures of Buckaroo Banzai.

ALIENS

20th-Century Fox (US), 1986, color, 137 min.

Credits: Producer, Gale Ann Hurd; director, James Cameron; story, James Cameron, David Giler, and Walter Hill; screenplay, James Cameron; special effects, John Richardson; music, James Horner.

Cast: Sigourney Weaver (Ripley), Michael Biehn (Corporal Hicks), Carrie Henn (Newt), Paul Reiser (Burks), Lance Henriksen (Bishop), William Hope (Lt. Gorman), Bill Paxton (Private Hudson), and Jenette Goldstein (Private Vasquez).

PLOT SUMMARY

At the start of the film a drifting space capsule is entered by boarders who are surprised to find that it contains an occupant, Ripley, in hibernation. After she is revived, Ripley is informed she has been asleep for 57 years. She is the sole survivor of the crew of a freighter that landed on a distant planet and unknowingly brought on board an alien creature found on a derelict spaceship. The creature killed all of the other crew members. Ripley then destroyed the freighter in an attempt to kill the alien, which managed to board her escape capsule before it left the doomed freighter. She ejected the alien from the capsule before going into a state of hibernation.

Ripley continues to have nightmares of the alien creature, which gestates inside a human host as part of its life cycle and then literally bursts from the host's chest.

The company which owned the freighter has in the intervening 57 years sent a group of colonists to terra-form the planet on which the alien was discovered by Ripley's crew. Contact is lost with the colonists and a company official, Burks, finally persuades Ripley to return to the planet with a contingent of space soldiers who are armed with state-of-the-art weaponry.

We next see Ripley, Burks, and the troopers awakening from hibernation as their spaceship nears the target planet. They move about as though they are in full gravity. Their commanding officer, Lt. Gorman, reviews what is known about the alien creatures. The soldiers refuse to believe Ripley's warnings that the aliens are extremely dangerous. Also among the military contingent is an android, Bishop. Ripley reacts with great hostility toward Bishop since another android, assigned to her space freighter, was responsible for bringing the alien on board because it had been secretly ordered by the company to return an alien specimen for study.

The entire contingent leaves the mother ship in a tactical reconnaissance space-ship that drops from the mother ship like an elevator whose cable has snapped. The spaceship lands near the colonists' settlement and a heavily armored vehicle proceeds toward the settlement. Two of the soldiers remain with the spaceship while the others proceed to search for the colonists, who have seemingly vanished. However, the expedition does find one survivor, a young girl, Newt.

The expedition learns that each colonist has an identification bracelet that can be located via the installation's computers. Sensors detect all of the bracelets in a huge facility housing a thermonuclear reactor. The expedition proceeds to the building, and everyone enters except Lt. Gorman, Ripley, Bishop, Newt, and Burks. As the soldiers near their target, Ripley observes that if they fire their weapons they may damage the thermonuclear reactor. Lt. Gorman, who has no combat experience, then orders the troopers to remove their magazines but still to proceed. The soldiers discover one of the colonists who is an unwilling host for one of the aliens and then they are attacked by a large number of the aliens. In the ensuing battle, a number of the soldiers are killed and the reactor is damaged. Lt. Gorman becomes paralyzed by indecision, but Ripley drives the armored vehicle deep inside the complex to rescue the surviving troops. The expedition then withdraws from the complex and radios the reconnaissance ship to pick them up. As the reconnaissance ship flies toward the troops an alien which had entered the ship kills both of the crew. The ship then crashes into the armored personnel carrier. This leaves several stranded soldiers, with very little ammunition, to face the aliens.

Newt, who had survived alone for a number of months, leads the expedition into the administration building. They seal the entrance to the complex and then discover, to their horror, that the damage done to the nuclear reactor will cause it to explode in a matter of hours. The android is sent via a long tunnel to a communications point in another part of the settlement, from which a signal can be transmitted to the mother ship to dispatch a second reconnaissance ship to rescue the survivors. While this is taking place, Ripley discovers that two of the alien creatures had been captured by the colonists and were housed in the administration building. These are relatively small, foot-long, creatures which attach themselves onto human hosts, gestate inside the hosts, and emerge later in the form that quickly develops into fully grown aliens. Burks wishes to take these two aliens back to the company for research purposes despite Ripley's objections. Burks later seals the door to the room in which Ripley and Newt are sleeping and releases the two aliens. Ripley and the little girl barely manage to escape from these aliens. They accuse Burks of attempted murder. The soldier in charge, Corporal Hicks, decides to summarily execute Burks, but the small group is attacked by a huge number of aliens before the sentence can be carried out. In the ensuing battle everyone is killed except Ripley, Hicks, and Newt. However, the little girl falls into a pool of water as they flee and is carried off by one of the aliens before Ripley can reach her.

Ripley and an injured Hicks are reunited with Bishop, who has successfully brought the other reconnaissance ship to the planet. Ripley is positive that Newt is still alive and can be reached before the thermonuclear reactor explodes. In an action-packed finish to the film, Ripley leaves the wounded Hicks with Bishop on the reconnaissance ship and blasts her way deep into the complex. After finding Newt, Ripley discovers the queen's chamber in which the eggs containing future aliens are laid. She destroys the eggs, and is pursued by the enraged alien queen.

While Ripley and Newt board the reconnaissance vehicle, the queen also enters the spacecraft. They fly away as the thermonuclear reactor explodes. When they land inside the mother ship, the queen attacks the four of them. Ripley, donning a robot lifting outfit, faces the huge alien in hand-to-hand combat, and finally flings it from the ship to its death.

As the film ends, Ripley, Newt, Hicks, and Bishop all enter hibernation for the trip back to Earth.

STAR TREK: "ARENA"

Paramount Pictures (US), 1967, color, 50 min.

Episode No. 19 of the television series, *Star Trek*—original airdate January 19, 1967.

Credits: Producer, Gene L. Coon; executive producer, Gene Roddenberry; director, Joseph Pevney; screenplay by Gene L. Coon from a story by Fredric Brown.

Cast: William Shatner (Capt. James T. Kirk), Leonard Nimoy (Mr. Spock), DeForest Kelley (Dr. Leonard "Bones" McCoy), James Doohan (Chief Engineer Montgomery "Scotty" Scott), George Takei (Sulu), Walter Koenig (Chekov), Nichelle Nichols (Commander Uhura), Sean Kenney (Helmsman DePaul), Grant Woods (Lt. Commander Kelowitz), and Vic Perrin (the Metron).

PLOT SUMMARY

Stardate 3045.6: crew members of the Enterprise beam down to one of Starfleet's outposts, Cestus III. Once there, they discover that the base has been completely destroyed. They are then attacked by an unseen enemy, as the Enterprise also does battle with an alien spaceship. The aliens retreat, and once back on board the Enterprise, Kirk decides to follow and destroy them in order to prevent future attacks on Starfleet bases. Spock is against destroying the aliens, but Kirk is convinced that they are evil and a threat to humanity.

The ensuing chase takes the Enterprise deep into uncharted space, as the ship is pushed to its maximum speed. Just as the Enterprise is about to overtake the enemy, both ships are immobilized by a mysterious force. The Enterprise crew then receives a message from the Metrons, an advanced race, who inform them that they and the aliens they are pursuing, the Gorns, have invaded the Metrons' section of the galaxy on an unacceptable "mission of violence." The Metrons decide to settle the conflict by sending the captains of both ships to the surface of an asteroid to fight to the death. The ship of the losing captain will also be destroyed, while the winner, and his ship, will be free to go. The Metrons tell Kirk that this kind of battle is appropriate for the "uncivilized" natures of the combatants. The battle will pit "ingenuity versus ingenuity, brute strength versus brute strength." The Metrons also inform Kirk that the asteroid has enough raw materials to enable each of the combatants to construct weapons capable of killing the other.

Kirk is then transported to the asteroid where he faces the alien captain, a huge reptilian creature who is much stronger than Kirk, but also much slower. The crew of the Enterprise cannot communicate with or help Kirk, but are forced to wait. While searching the asteroid's barren surface for the materials with which to construct weapons, Kirk finds diamonds, and muses that, despite their value, he "would trade them all for a hand phaser or a good solid club" to use to fight the

Gorn. He also says, finding deposits of sulfur, that the planet is "a mineralogist's dream" but he is increasingly worried because he cannot find any kind of weapon, which he desperately needs to kill the Gorn.

Kirk manages to injure the Gorn by rolling a boulder down a hill onto the creature, but the Gorn recovers. Soon Kirk falls into a trap set by the Gorn, injuring his leg. Although Kirk is able to hobble away, he grows weaker, and the Metrons contact the Enterprise's bridge crew to tell them that their captain is losing the battle.

The Metrons then allow the crew to see the action on the asteroid. The Gorn speaks to Kirk, informing him that Starfleet had invaded the Gorn's territory by establishing Cestus III, and that was the reason for the destruction of the outpost by the Gorns.

Kirk finds a deposit of potassium nitrate, which seems to give him an idea. Spock, watching from the Enterprise bridge, also reaches the same conclusion: Kirk can build a primitive gun— using potassium nitrate, sulfur, and coal to make gunpowder; diamonds as projectiles; and a piece of hollow wood as the barrel. Kirk gathers the necessary components and races against time to construct his weapon before the Gorn reaches him. Kirk makes the gunpowder and loads the primitive gun. He creates a spark by rubbing his communicator against a rock, igniting the gunpowder just as the Gorn approaches him. The Gorn is stunned but not killed. Kirk picks up the Gorn's dagger and is about to kill the creature with it when he stops, saying that the Gorns' destruction of Cestus III may have been justified as a defense of their territory against invaders.

Kirk shouts that he will not kill the Gorn. Suddenly the Gorn vanishes and a Metron, a boyish creature in a shimmering toga, materializes. The Metron says that Kirk's act shows "the advanced trait of mercy" and that there is hope for the human race to become truly civilized. The Metron sends Kirk back to the Enterprise, and the Enterprise is suddenly transported across the galaxy, ready for new adventures.

THE BLACK HOLE

Walt Disney (US), 1979, color, 97 min.

Credits: Producer, Ron Miller; director, Gary Nelson; screenplay, Jeb Rosebrook and Gerry Day, from a story by Jeb Rosebrook, Bob Barbash, and Richard Landau; cinematographer, Frank Phillips; music, John Barry.

Cast: Maximillian Schell (Dr. Hans Reinhardt), Anthony Perkins (Dr. Alex Durant), Robert Forster (Captain Dan Holland), Joseph Bottoms (Lt. Charles Pizer), Yvette Mimieux (Dr. Kate McCrae), and Ernest Borgnine (Harry Booth).

PLOT SUMMARY

In the year 2130, five astronauts are on a mission to search for "intelligent life" in the galaxy. As the film opens, their ship, the Palomino, is slightly off course. Their robot assistant, Vincent, discovers that an enormous black hole is the cause of this deviation. The astronauts gather to observe the black hole, and soon see a stationary ship that is located very near to it. By analyzing the ship's design, they conclude that it is the Cygnus, an American ship lost decades earlier. Crew member Dr. Kate McCrae's father was aboard the Cygnus when it disappeared.

The astronauts discuss the reputation of the ship's captain, Dr. Hans Reinhardt, and agree that he had been both a genius and a charismatic individual who was able to convince Congress to fund his scientific interests as though they were a matter of national priority. They decide to examine the Cygnus at closer range despite the danger of coming so close to the black hole. As they approach the Cygnus, suddenly the powerful gravitational pull of the black hole, which had been stressing their ship, disappears. Four of the astronauts and the robot board the Cygnus. Their weapons are soon shot out of their hands by unseen assailants, and they are led deeper into the ship by doors that open and close mysteriously behind them. Finally they are brought to the control tower of the ship, where they meet the ship's captain, Dr. Hans Reinhardt.

Dr. McCrae asks him about her father's fate. Reinhardt tells her that her father is dead along with the other members of the original crew who perished when their ship was lost on a return voyage to Earth. Reinhardt states that he has created "companions of a sort"—robots that operate the ship and carry out his orders. Reinhardt seems hospitable to the Palomino crew, and offers to supply them with the materials needed to repair the damage incurred by their ship as they approached the Cygnus.

Reinhardt tells the Palomino crew that he chose not to return to Earth (disobeying a direct command to do so) because he wanted to continue his work on developing a power source that would enable him to navigate through the black hole. It is this power source that permits him to maintain the stationary position of the Cygnus near the opening to the black hole. Reinhardt believes that traveling to the other side of the black hole will give him unparalleled knowledge about the universe.

Some of the crew members explore the Cygnus on their own and discover such oddities as evidence of a robot's funeral and a limping robot. Meanwhile, their robot, Vincent, befriends a similar robot, Bob, on the Cygnus. Bob states that the robot crew is actually the original human crew who have been turned into automatons. Vincent and Bob call back three of the Palomino crew to their ship, and tell them about the humans/robots. The astronauts then ask their shipmates, Dr. Durant and Dr. McCrae, who are still on board the Cygnus, to return at once to the Palomino. Vincent and Dr. McCrae are able to communicate telepathically. When McCrae is unable to convince Durant to leave the Cygnus, Vincent tells her about the humans/robots who operate the Cygnus. McCrae is very upset about her father's fate but conveys this information to Durant, who then unmasks one of the "robots." Then Durant and McCrae try to escape.

Maximillian, the leader of Reinhardt's real robot crew, kills Durant, and Reinhardt orders his robots to take McCrae to the "hospital," where she will also be transformed into a robot/human. Vincent tells the Palomino crew what has happened and they go to rescue Dr. McCrae. Booth, a coward, returns to the Palomino on a pretense and attempts to flee, but the ship is destroyed by Reinhardt. McCrae is rescued, but the crew are now stranded on the Cygnus. As they try to reach the Cygnus' probe ship, a meteor storm begins to buffet the Cygnus and the ship begins to disintegrate. In the control tower, Dr. Reinhardt is crushed by fallen debris while the Palomino crew has a shoot out with his robot army. As the Palomino crew reaches the probe ship, Maximillian attacks them. Vincent overcomes Maximillian, and the crew enters the shuttle craft. As they leave the Cygnus, they realize that the probe ship is preprogrammed to go through the black hole. The film ends with their safe passage through the black hole.

THE EMPIRE STRIKES BACK

Twentieth-Century Fox (US), 1980, color, 124 min.

Credits: Producer, Gary Kurtz; executive producer, George Lucas; director, Irvin Kershner, screenplay by Leigh Brackett and Lawrence Kasdan from a story by George Lucas; cinematographer, Peter Suschitzy; music, John Williams; special video effects by Brian Johnson and Richard Edlund.

Cast: Mark Hamill (Luke Skywalker), Harrison Ford (Han Solo), Carrie Fisher (Princess Leia Organa), Billy Dee Williams (Lando Calrissian), Anthony Daniels (C-3PO), David Prowse (Darth Vader), Peter Mayhew (Chewbacca), Kenny Baker (R2-D2), Frank Oz (Yoda), Alec Guinness (Obi-Wan Kenobi), and Jeremy Bulloch (Boba Fett).

PLOT SUMMARY

The film begins with a Star Wars-like prologue in which the audience learns that the struggle between the Rebels and the Imperial forces continues. The freedom fighters are secretly stationed on the ice planet, Hoth, while the evil Empire is searching tirelessly for them. Darth Vader has sent thousands of probes into space in an attempt to find them.

The film then shifts to the planet Hoth, where one of the Empire's space probes has landed. Luke Skywalker is out exploring: he is attacked by an ice beast, who drags him away. Meanwhile, back at the Rebel base, Han Solo is preparing to leave in order to repay his debt to Jabba the Hut. Solo banters with Leia, suggesting that she is in love with him, which she vehemently denies.

As nightfall approaches and Skywalker has still not returned, Solo ventures out alone to find him, and succeeds in doing so just as his mount dies from the extreme cold. Solo constructs a shelter which protects them until they are rescued the next day. Shortly thereafter the Imperial forces begin attacking the rebel base. The Rebels engage in a ground battle against Imperial walkers, enormous tanks on stilts. Meanwhile, Rebel fighter planes escort the transport ships off the planet to another rendezvous point. Solo, Leia, Chewbacca, and C-3PO manage to escape on Solo's ship, the *Millennium Falcon*, but are forced to enter an asteroid field when the *Falcon's* hyperdrive (which enables it to move at light speed) malfunctions. Solo is able to navigate the ship into a cave where they hide from the Imperial forces.

Skywalker decides not to meet the others at the Rebel rendezvous point. Instead he goes to the Dagobah System (to which he has been directed by the voice of Ben Kenobi), in order to study under the Jedi master, Yoda. When Skywalker arrives at Dagobah, his ship sinks into a swamp. He soon finds a tiny, odd, elflike creature who promises to lead the impatient young man to Yoda. Finally, the creature reveals itself to be Yoda, but expresses doubt to Obi-Wan Kenobi's ghost about Skywalker's readiness to become a Jedi Knight.

Back on the *Millennium Falcon*, Solo and Leia discover that they are strongly attracted to each other, and also that the cave they are in is really the throat of a gigantic space beast. They manage to escape in the nick of time. Meanwhile, Darth Vader tells the evil Emperor that Skywalker would be a great asset to the Dark Side (the evil component of the mysterious "force," which the Jedi Knights use for good) and vows that Skywalker will join with the Empire or die.

Now no longer hidden, the *Falcon* is pursued by an Imperial warship, but manages to hide again by attaching itself to the underside of the Imperial ship. The *Falcon* floats away when the warship jettisons its garbage into space. As Solo and Chewbacca are still unable to repair the *Falcon's* hyperdrive, Solo decides to fly to the Bespin System and visit a small mining colony that is run by an old friend of his, Lando Calrissian. They do not notice that a ship is following them, piloted by the bounty hunter Boba Fett.

Back on Dagobah, Skywalker is learning from Yoda, but he is still far from completing his training. He senses the suffering of his friends, and when Yoda tells him that it is the future he feels, Skywalker decides he must go to help them, despite warnings from Yoda and Kenobi that he is not yet powerful enough to defeat Darth Vader.

Not completely trusting Lando upon their arrival at Bespin, Solo, Leia, Chewbacca, and C-3PO soon discover that Lando has indeed betrayed them to the Empire. They are tortured and surmise that they are being used as bait to lure Skywalker to Vader. Just before Skywalker arrives, Vader tests the carbon freeze system on Solo. He intends to use it later on Skywalker. Vader then turns the hibernating Solo over to Boba Fett, who plans to take him to Jabba the Hut for the bounty that Jabba has offered for Solo.

Skywalker lands and soon is face to face with Darth Vader. They battle, and it becomes obvious that Vader is much more powerful than Skywalker. Vader tries to convince Skywalker to join the Dark Side, but Skywalker resists. Vader chops off Skywalker's right hand in combat, and then tells him that he is Skywalker's father. Not wanting to believe this, Skywalker lets himself fall down a long shaft, where he clings desperately to a pipe, close to death. Meanwhile, Lando has helped Leia, Chewbacca, and the droids to escape on the *Millennium Falcon*. Leia "hears" Skywalker's cry for help and directs them to his location, where they are able to save him. In the nick of time, R2-D2 reactivates the *Falcon's* hyperdrive and they escape from the clutches of the Empire. The film ends with Lando and Chewbacca going off to search for Solo as Skywalker and Leia watch from a window of a rebel spaceship.

THE FLY

Twentieth-Century Fox (US), 1986, color, 96 min.

Credits: Producer, Stuart Kornfeld; director, David Cronenberg; screenplay, Charles Edward Pogue and David Cronenberg based on a story by George Langelaan; cinematographer, Mark Irwin; music, Howard Shore; special effects, Louis Craig; supervisor of computer/video effects, Lee Wilson.

Cast: Jeff Goldblum (Seth Brundle), Geena Davis (Veronica Quaife), John Getz (Stathis Borans), Joy Boushel (Tawny), Les Carlson (Dr. Cheevers), George Chuvalo (Marky), and David Cronenberg (Gynecologist).

PLOT SUMMARY

The film opens with Veronica Quaife, a reporter for a science magazine, and Seth Brundle, a strange-acting scientist, meeting at a party. Brundle clumsily tries to pick up Quaife by dropping hints about a project he is working on that "will change life as we know it." Claiming he does not want to discuss his secret project in public, he lures her back to his loft/laboratory to give her a demonstration. Once there, he shows off the human-sized "telepods" he has designed to transport matter—an object can be placed in one capsule, disintegrated, and be reintegrated in another capsule across the room. Brundle demonstrates by transporting one of Davis' stockings, but convinces her not to write about his invention yet, as he is thus far only able to transport inanimate objects. Quaife agrees, but talks to her editor (and former lover), Stathis Borans, about the invention. Borans is both suspicious of Brundle's invention and jealous of Quaife's interest in the scientist.

As Brundle continues to perfect his invention, he and Quaife fall in love. Despairing after a failed attempt at transporting a baboon (which results in the baboon being turned inside out), he gets an idea from Quaife's off-hand comment about flesh and is soon able to transport a baboon successfully. Later that evening, Quaife leaves Brundle's loft to finalize her break-up with Borans. Lonely, Brundle gets drunk and decides to transport himself in the machine. Unfortunately, a housefly is also inside the pod with Brundle, and the computer's reintegration program fuses the scientist and the fly into one being. After his transportation, Brundle finds himself with incredible appetite, strength and virility, and attributes this "purification" to the transportation he has undergone.

He urges Quaife to go through the process, but she refuses, alarmed by his sudden change, not only physically, but in personality as well. She notices that long dark hairs are growing on his back and takes some to be analyzed. She discovers that the hairs are insectlike, and soon what has happened becomes painfully obvious. Over the next few weeks, Brundle is able to keep his sarcastic sense of humor as he becomes dramatically more and more flylike—his appendages begin to fall off, he can walk on walls, he eats by spewing acid on his food, etc. Quaife discovers that she is pregnant, and becomes desperate to have an abortion because she is afraid that she was impregnated after Brundle's fusion. Thus the unborn child will be part fly.

Brundle learns of her decision to have an abortion and carries her back to his laboratory. He then tries to fuse her together with him in the pods. Borans follows them back to the laboratory and saves Quaife by severing the power cord to her pod. As a result, Brundle, now almost completely fly, is fused together with one of the pods. As the lump of fly/metal crawls out of the other pod, Quaife mercifully kills it.

METEOR

American International (US), 1979, color, 103 min.

Credits: Producer, Arnold Orgolini and Theodore Parvin; director, Ronald Neame; screenplay, Stanley Mann and Edmund H. North from a story by North; cinematographer, Paul Lohmann; music, Laurence Rosenthal; special effects, Glen Robinson.

Cast: Sean Connery (Dr. Paul Bradley), Natalie Wood (Tatiana Nikolaevna Donskaya), Brian Keith (Dr. Alexei Dubov), Martin Landau (Gen. Barry Adlon), Trevor Howard (Sir Michael Hughes), Joseph Campanella (Gen. Easton), Roger Robinson (Hunter), Karl Malden (Harry Sherwood), Richard A. Dysart (Secretary of Defense), and Henry Fonda (The President of the US).

PLOT SUMMARY

The film begins with a voice-over narration about "outer space, limitless and timeless" that discusses comets, asteroid belts, and describes a giant asteroid, Orpheus, 20 miles in diameter and sitting in an asteroid belt in our solar system, that has "never been disturbed until now... ." The voice goes on to say that a comet is heading directly toward the asteroid belt, threatening to send meteors flying through space.

In the next scene, a Coast Guard ship interrupts a sailboat race to pick up an American scientist, Dr. Paul Bradley, who is transported to Washington, where he arrives at Harry Sherwood's office. Sherwood, a NASA official and Bradley's former boss, tells him that a comet has collided with the gigantic asteroid Orpheus. The explosion has sent a rock five miles in diameter hurtling toward Earth. It is due to strike the Earth in a few days' time. Sherwood asks Bradley to accompany him to a Washington meeting to decide what to do.

The audience learns that Bradley had designed an orbiting Star Wars-like space defense platform with warheads that would face into outer space. The orbiting platform, however, has its warheads aimed at the Soviet Union and China. Bradley agrees to help realign the warheads in order to destroy the huge rock heading toward Earth. At the Washington meeting, it becomes apparent that Hercules, as the project is code-named, is a breach of an international treaty. Thus, government officials are loathe to acknowledge its existence. Bradley describes the horrifying damage, including another Ice Age, that could occur if the meteor hits the Earth. He also tells the President that Hercules' rockets are insufficient by themselves to destroy the meteor. The task will also require the rockets of a similar Soviet space platform.

The President gives a public address telling Americans that Hercules was developed with exactly this kind of crisis in mind, and that the American and Soviet governments are going to cooperate. The Russians are invited to come to the US to work on the problem. We see Russian governmental leaders watching the address; they remark that the President "has turned hypocrisy into diplomacy."

The next day, Bradley and Sherwood go to Hercules' secret headquarters below a New York City skyscraper. Against the wishes of the American general formerly heading the Hercules project, a Russian scientist, Dubov, arrives at the secret headquarters with his assistant/interpreter, Donskaya. A mutual attraction develops between Donskaya and Bradley. Realizing the seriousness of the situation, the Russians quickly admit to their own system, called Peter the Great, and the two teams begin working together to solve the problem.

Meanwhile, smaller meteors, called "splinters," begin hitting the Earth, causing an earthquake in Siberia, an avalanche in the Swiss Alps, and a tidal wave in Hong Kong. The American and Russian scientists time the two sets of missiles to hit the meteor simultaneously. Before Hercules can be fired, however, the scientists discover that a giant splinter is heading directly towards New York City. Luckily, they are able to set off Hercules just before the splinter hits New York.

The impact of the splinter traps the scientists underground. They are able to escape through a subway tunnel, and they then learn that Peter the Great and Hercules have destroyed the meteor. The film ends with the Russian scientists leaving, and a footnote tells us of a real-life project at the Massachusetts Institute of Technology, called Icarus, designed to destroy a meteor on a collision course with Earth.

PHASE IV

Paramount Pictures (US), 1974, color, 84 min.

Credits: Producer, Paul B. Radin; director, Saul Bass; screenplay, Mayo Simon; cinematography, Dick Bush and Ken Middleham; special effects, John Richardson; music, Brian Gascoigne and Yamashta.

Cast: Nigel Davenport (Ernest Hubbs), Michael Murphy (James Lesko), Lynne Frederick (Kendra), Alan Gifford (Mr. Eldridge), Robert Henderson (Clete), and Helen Horton (Mrs. Eldridge).

PLOT SUMMARY

The film begins with a prologue describing how "that spring, everyone was watching the skies"—and we see what looks like a solar eclipse and unusual displays of radiation in space. The narrator tells us that these events in space have caused effects on Earth, which have gone almost unnoticed, except by one biologist, Dr. Ernest Hubbs, who has investigated a new phenomenon: ants of different species are communicating with each other.

Hubbs convinces the National Science Foundation to give him funding to explore a "biological imbalance" in the Arizona desert. Apparently the ants of the area have been successful in killing their predators, resulting in a huge increase in the ant population. Hubbs proposes to create an observation/research center to study the ants and control them. He brings a cryptographer, James Lesko, with him, hoping that Lesko will be able to decipher the ants' communication system.

Hubbs and Lesko survey the immediate area and find a demolished house and strange towers made of sand, which they cannot explain. The pair warn the local residents to evacuate, but one family ignores the warnings. When the ants invade their farm, the family flees to the nearby research center. Unaware that the humans are approaching, the scientists release a poison into the air, killing both the ants and the fleeing humans. A young girl survives by hiding in a shed next to the research facility, however, and she is rescued by the researchers when they go outside the next morning to collect ant specimens. Inside the center's lab, the girl, Kendra, destroys the researcher's ant tank in a fit of anger, releasing the ants. One of the ants bites Hubbs.

Meanwhile, outside the center, the ants who have survived the poison have created sand structures that slant toward the research center reflecting sunlight into it. This causes the temperature inside the research center to rise dramatically. The scientists are able to destroy the structures using sound waves, but the ants inside the center manage to disable its air conditioning system, causing its computers to malfunction. The ants also cutoff the center's communication with the outside world. Lesko sends the ants a simple message, and they reply with a message that seems to imply that they want one of the three humans.

Unnoticed, Kendra goes outside alone, believing that her sacrifice will save the others. Hubbs, deathly ill from the ant bite, decides that the queen ant must be killed, and also goes outside. He falls into a trap set by the ants and is devoured. Lesko then decides he must find the queen, but upon arrival at the ant's main tunnel, discovers Kendra. The reunited pair are somehow "changed" by the ants, and the film ends with Lesko and Kendra looking out towards the horizon, implying that they and the ants will conquer the rest of the world.

PLANET OF THE APES

20th-Century Fox (US), 1968, color, 112 min.

Credits: Producer, Arthur P. Jacobs; director, Franklin J. Schaffner; screenplay, Michael Wilson and Rod Serling, based on the novel *Monkey Planet* by Pierre Boulle; music, Jerry Goldsmith.

Cast: Charlton Heston (George Taylor), Linda Harrison (Nova), Roddy McDowall (Dr. Cornelius), Kim Hunter (Dr. Zira), Maurice Evans (Dr. Zaius), James Whitmore (President of the Assembly), James Daley (Honorious), and Robert Gunner (Landon).

PLOT SUMMARY

The film opens with astronaut George Taylor completing his final report before returning to Earth. While he and his crew have aged only six months, their ship's computer indicates that 700 years have elapsed on Earth because the ship was traveling at nearly the speed of light. At the conclusion of his message, Taylor gives himself an injection and straps himself into a sleep chamber like those occupied by the other three astronauts.

The spaceship lands on an unknown planet, in an area that resembles the Grand Canyon. The sleep chambers open, and three of the four astronauts emerge. The fourth astronaut, the only woman in the group, has died of old age because of an air leak. The ship's computer indicates that the year is now 3978 on Earth.

The three astronauts escape on a lifeboat as water floods the cabin of the spaceship. As they approach the shore, the spaceship sinks. After some discussion, they conclude that they do not know where they are. All they have been able to carry from the spaceship is a soil test kit, medical supplies, and enough food for a few days.

They explore the area for food and civilization, but encounter only a dangerous avalanche. By this time they have only 8 oz of water, and there is no sign of rain, although they see lightning and hear thunder. Suddenly, they discover life—a plant with bright yellow flowers. Their hope renewed, the men continue searching through the night and on the following day they notice people looking at them from atop a hill. These "people" turn out to be scarecrows made to resemble people. Beyond the hill they find a waterfall. They bathe and drink, but when they return for their clothes they find that most of them are missing. In the distance they see a tribe of primitive people. It appears that these people are unable to speak. When the tribe sees the astronauts, they panic and scatter.

Suddenly, gunshots are fired by horsemen. The astronauts are shocked when they realize that the gunmen are actually human-size *apes*. The apes beat and capture many of the humans. In the process one of the astronauts is mistaken for a primitive human and is killed by the apes. Taylor, shot in the throat, is captured along with the third astronaut. Taylor is surprised to see the apes taking pictures of themselves with their human kills, and to hear the apes speaking English. Taylor is placed in a cell in a research institute. Later a young woman is also put into Taylor's cell. He calls her "Nova."

He surmises that evolution has taken a different course on this planet. Humans cannot talk. They can be taught a few tricks but are basically wild and untamed. The apes regard humans as a nuisance. Dr Zaius, the simian Minister of Science

Planet of the Apes. George Taylor is held captive with other humans by a race of intelligent apes.
(Photo: Museum of Modern Art/Film Stills Archive. Courtesy of 20th Century-Fox.)

observes to his colleagues that human scavengers ravage their crops and the sooner they are exterminated, the better.

Zaius' hatred is prompted by an ulterior motive: he has secretly discovered that humans once ruled their planet. This fact is at odds with ape history as expounded in the Simian Bible, called the Twelve Scrolls. If the truth were known, the foundation of ape culture would be undermined. Zaius, who is also Defender of the Faith, feels obligated to suppress this information.

Dr. Zira, a researcher, and her fiance, Dr. Cornelius, an archaeologist, discover that Taylor is different than other humans. Because of his gunshot wound, Taylor temporarily cannot talk, but he demonstrates his ability to understand Zira by writing in the sand and then by handing her a note stating that his name is "Taylor." Taylor tells Cornelius and Zira that his spaceship landed in an uninhabited area which the apes refer to as the Forbidden Zone. Cornelius is excited because he has unearthed evidence of an older human civilization in this area. He has not been able to convince anyone, but Taylor appears to be a living example of his discovery.

When Zaius learns of Taylor's verbal abilities, he puts him on trial to dispose of him and threatens to accuse Zira and Cornelius of scientific heresy if they aid in Taylor's defense. In the course of the trial Taylor discovers that a lobotomy has been performed on the third astronaut to prevent him from also demonstrating the ability of speech. Secretly Zira and Cornelius help Taylor to flee to the excavation site. When Zaius goes there, he is confronted by one of their findings—the remains of a talking doll from an older civilization of humans. Zaius still will not concede. He asks, "Why, if man were superior, did he not survive?" No one can answer that question.

The question is answered when Taylor and Nova win their freedom and ride off on horseback along the deserted shore. Ahead, Taylor sees a massive structure which turns out to be the Statue of Liberty, now almost completely covered by sand. He now realizes that his journey has taken him forward in time to the Earth. In a rage, he yells, "You finally did it! You murderers! You blew it up!"

THE PRODUCTION AND DECAY OF STRANGE PARTICLES (THE OUTER LIMITS)

Daystar-United Artists (US), 1964, black and white, 52 min.

Credits: Producer, Joseph Stefano; director and screenplay writer, Leslie Stevens; special effects, Si Simonson.

Cast: George Macready (Dr. Marshall), Robert Foutier (Dr. Pollard), Leonard Nimoy (Koenig), and Signe Hasso (Laurel Marshall).

PLOT SUMMARY

The program opens with the narrator explaining that, recently, subatomic particles have been discovered. These are fragments of atomic particles, smaller than the smallest atom. They include antimatter, made up of "inside out" material, and shadow matter, a strange material which is powerful enough to penetrate many inches of lead.

The story takes place in a nuclear reactor facility which is investigating a new isotope. In order to protect themselves while handling radioactive materials, the technicians use special gloves with metal fingers and stand behind a thick wall. They also wear special shielded suits to further reduce the radiation reaching them.

Something has gone wrong while working with this new isotope. The work team has been exposed to dangerous levels of radiation and thus they are ordered to leave the research room immediately.

One of the men, Collins, handling the new isotope with his gloves, is unable to release it. We later see that he has been transformed into an energy creature: only electric discharges can be seen inside his helmet. Another worker, Koenig, goes to Collins' assistance but is also transformed by this mysterious isotope. Twenty feet of lead slabs are put up to shield the staff from the radiation, but the radiation which penetrates the slabs continues to increase.

The head of the facility, Dr. Marshall, fears that there will be a chain reaction although he cannot understand how the reactor fuel can form a critical mass with the control rods pushed in. There seems to be some sort of spontaneous generation of radiation.

The scientists' wives then arrive and chide them for not appearing at the races. When the wives learn that there has been an emergency, they insist on staying with their husbands until everyone leaves the facility.

The technicians whose bodies have been transformed into "particle creatures" by the radiation extend their power by touching each other as the unaffected staff place more shields to reduce the radiation levels. One man putting up a shield is attacked by Collins. He reports that Collins' face was only a bright blue light, like an electric arc lamp. The energy emitted through Collins' visor burned the man's face and blinded him; he soon dies.

As the reactor approaches the critical mass which should ignite a chain reaction, Marshall becomes paralyzed with fear. He tells his wife Laurel, that his experiments with heavy elements somehow caused this and that now the radiation is growing geometrically. There is no hope. She implores him to try to solve the crisis. Marshall finally gets up the courage to go into the lab. He observes that the particle creatures dissolve when they pass behind thick radiation shields that separate them from their energy source.

Marshall at first cannot understand what is delaying a fission explosion since the substance has now grown to larger than the critical mass needed to produce such an event. He surmises that the energy released by the splitting nuclei is producing tiny cracks into another dimension. He believes that he can counteract this by setting up a fusion reaction.

Marshall puts on a special radiation protection suit and constructs a hydrogen bomb component, which he leaves for the "particle creatures." The facility is evacuated and Marshall and Laurel watch the ensuing explosion from a bomb shelter several miles away. There is a nuclear blast, followed by a tremendous wind storm. Suddenly, the matter implodes as time flows backwards. Buildings which were blown away are now standing, and all is as it was before the explosion.

At the end the narrator states that humanity can use its scientific discoveries for constructive or destructive purposes...a Cold War warning.

SILENT RUNNING

Universal Pictures (US), 1971, color, 90 min.

Credits: Producer, Michael Gruskoff; director, Douglas Trumbull; screenplay, Deric Washburn, Mike Cimino, and Steven Bochco; original songs sung by Joan Baez; music, Peter Schickele.

Cast: Bruce Dern (Lowell), Cliff Potts (Wolf), Ron Rifkin (Barker), Jesse Vint (Keenan), and Mark Persons, Steven Brown, Cheryl Sparks, and Larry Wisenhunt (Drones).

PLOT SUMMARY

The film opens with beautiful close-ups of flowers, lush plants, snails, frogs, and other small animals. Astronaut Lowell swims in a clear, rapidly flowing stream. He gets out and talks to some rabbits who are watching him. Suddenly, we realize that he is not outdoors but inside a gigantic glass dome that is in orbit about Saturn.

He has three shipmates who have no respect for him or his garden. The crew also includes three little robots, or drones, who help maintain the space station. Their whereabouts and activities are monitored through televisions.

The four shipmates gather around the kitchen table and discuss their mission. They have been sent into space aboard the *Valley Forge*, which, together with other spaceships, contain the world's last remaining forests. They are the keepers of these space greenhouses. Lowell's colleagues eat only synthetic food, while he consumes real vegetables from the domes.

There are several craft similar to the *Valley Forge,* each carrying several glass-dome-enclosed forests. While these craft are onscreen the film repeats the speech that was given when these forests were dedicated and sent into space. It was hoped that the forests would some day be returned to Earth to reforest the planet.

One day, the crew is ordered to abandon the project and immediately nuclear destruct the forests so that the spacecraft can revert back to commercial freighters. While his three colleagues celebrate their impending return to Earth, a depressed Lowell goes into the forest and broods. When he returns to the galley, he becomes enraged when he sees his friends eating their synthetic food. Soon this will be the only form of sustenance left. He tries to persuade them to appreciate the difference between the synthetic foods they eat and the flavorful natural grown foods. "On Earth," he says, "the temperature is always 75°. All people are the same. Everyone has a job. There's no beauty or frontiers." He laments that a little girl, whose photograph is on the wall, will never have the opportunity to touch a leaf. He begs the rangers not to blow up the domes. They refuse.

The other three crewmen go off to the forest to detonate the bombs and one of them is injured. While Lowell treats the injured crewman, he hears over the intercom that bombs will detonate the forests from the other ships in 5 minutes. Then we see these domes explode. Lowell confronts one of his coworkers who is

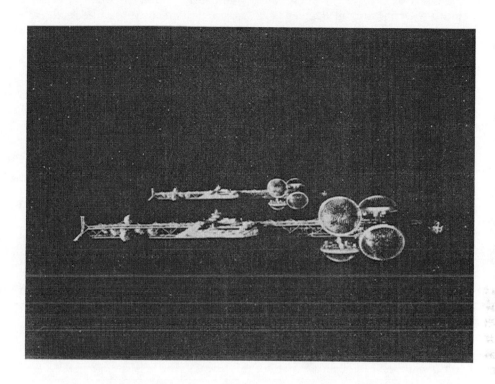

Silent Running. Space freighters carrying glass-dome enclosed forests in orbit around Saturn. (Photo: Museum of Modern Art/Film Stills Archive. Courtesy of Universal Pictures.)

trying to arm a bomb in the dome in which Lowell is working. In the fight that ensues, Lowell kills the man, but injures his own leg. Subsequently he jettisons the dome on which the remaining two workers are arming another bomb, and detonates the bomb, killing them.

Lowell makes a tourniquet out of an electric cable to staunch the bleeding of his leg, and informs mission control that he is having trouble with *Valley Forge's* main couplings. Realizing that he cannot return to Earth with the other ships, he steers the *Valley Forge* toward Saturn and reports that an onboard accident has damaged the ship and killed all of his crewmates.

He programs the drones to operate on his badly injured leg. After he awakens from the surgery, mission control tells him that if he is unable to steer the damaged craft away from Saturn's rings he will die. They suggest that he consider suicide.

As Lowell sleeps, the *Valley Forge* passes through Saturn's rings. The spacecraft's violent shaking awakens Lowell, who instructs the drones to return from outside of the ship. Drone Three becomes disoriented and is blown away, but the ship survives the passage through the rings.

Mission control believes that the *Valley Forge* has been destroyed. Thus Lowell is entirely on his own. Missing human interaction, he names one drone "Huey" and the second drone "Dewey." He programs them to respond directly to him and to devote more time to maintaining the forest.

Lowell, alone in space, becomes depressed. To fight his loneliness, he programs the drones to play poker. They secretly cheat by showing each other their cards and by communicating with one another.

The forest then begins to wilt. Distressed that his beloved forest is dying, Lowell accidentally runs over Huey. Lowell attempts to fix Huey with the available parts onboard but cannot make adequate repairs. Huey, who seems to possess almost human feelings, is left a cripple.

To make matters worse, mission control contacts Lowell and congratulates him for surviving his journey through Saturn's rings. They have located the *Valley Forge* and are sending a rescue team. They tell him that the ship was hard to find because it is extremely dark where it is located. Only then does Lowell realize that the deterioration of the forest is the result of a lack of sunlight.

Lowell installs a powerful artificial light system for the plants. Knowing that he will soon be rescued, he tells Dewey to stay in the dome and maintain the forest. Lowell separates the dome from the Valley Forge. He then detonates a bomb, destroying the *Valley Forge*, Huey, and himself. The dome's forest, however, remains intact and the film ends with Dewey obediently tending it.

STAR TREK IV: THE VOYAGE HOME

Paramount Pictures (US), 1986, color, 119 min.

Credits: Producer, Harve Bennett; director, Leonard Nimoy; screenplay, Harve Bennett, Steve Meerson, Peter Krikes, and Nicholas Meyer, based on a story by Leonard Nimoy and Harve Bennett and the TV series created by Gene Roddenberry; cinematographer, Don Peterman; special effects, Michael Lanteri; music, Leonard Rosenman.

Cast: William Shatner (Admiral James T. Kirk), Leonard Nimoy (Mr. Spock), DeForest Kelley (Dr. Leonard "Bones" McCoy), James Doohan (Chief Engineer Montgomery "Scotty" Scott), George Takei (Sulu), Walter Koenig (Chekov), Nichelle Nichols (Commander Uhura), Jane Wyatt (Amanda, Spock's mother), and Catherine Hicks (Dr. Gillian Taylor).

PLOT SUMMARY

The film opens with the crew of the *Enterprise* facing charges, in absentia, of mutiny, destroying property, and various other crimes. The accuser is Schuck, a Klingon representative, who wants the *Enterprise* crew extradited to the Klingon Empire for having killed a Klingon in the previous film episode, *Star Trek III*.

The crew of the *Enterprise* then returns to Earth in a captured Klingon spaceship to face these charges when their sensors detect a strange sound emanating from a huge space probe that is near Earth. The space probe had earlier neutralized the energy sources in every spaceship it encountered on its way to Earth.

Captain Kirk and his crew conclude that the space probe is emulating the cry of the now-extinct hump-backed whale, and if it is unsuccessful in contacting a whale, will destroy Earth. In order to save the planet, they must time travel to the 20th century and bring hump-backed whales back to the future.

The crew travels back in time by accelerating to extremely high speeds, using the gravitational field of the Sun. They land in San Francisco, in 1986, and use their ship's cloaking device to hide it in a park. Although they are dressed anachronistically, no one seems to notice them as unusual. The crew divides itself into three groups to accomplish three tasks: finding whales, building a whale tank, and locating a power source to repair their ship's power plant, which was damaged in the journey back in time.

Chekov and Uhura go to the harbor to look for a nuclear fission source aboard a nuclear vessel, and are immediately mistaken for spies by naval security, who capture Chekov. When he subsequently tries to escape, he is badly injured and is taken to a hospital. The *Enterprise* crew manage to sneak into the hospital and McCoy saves his friend's life.

Earlier, Scott and McCoy searched for materials with which to build the needed whale tank, and gave a 20th-century engineer some 23rd-century technology in order to accomplish their goal.

Kirk and Spock visit a local aquarium which houses a pair of whales, George and Gracie, who are in the care of Dr. Gillian Taylor. She is being forced to release the whales into the sea because of the cost of feeding them, and she is worried that they will fall prey to hunters once they are released from captivity. Spock dives into George and Gracie's tank to converse with them, and he and Kirk are promptly ushered out of the aquarium by Dr. Taylor. Later, they again encounter Taylor, who is stunned when Spock tells her that he knows that Gracie is pregnant. Kirk takes Taylor to dinner to try to enlist her support, and only their mutual attraction keeps Taylor from leaving after hearing Kirk's unbelievable story.

Later, when the whales are released into the sea, the *Enterprise,* with Taylor on board, goes after them. The crew manages to beam George and Gracie into their ship's now-finished whale tank just seconds before the whales' almost certain death by hunters' harpoons. The crew and whales then return to the 23rd century, accompanied by Taylor. When the space probe hears the whales respond to its call, it leaves our solar system and the Earth is spared.

At the end of the film, Kirk and the crew face their tribunal, and are cleared of all charges except one: Kirk's insubordination. For this, he is demoted from Admiral to Captain so that he and his crew can again pilot a starship. The crew happily exits, anticipating many new adventures.

STAR WARS

Twentieth-Century Fox (US), 1977, color, 119 min.

Credits: Producer, Gary Kurtz for Lucasfilm Productions; director and screenplay, George Lucas; cinematographer, Gilbert Taylor; special effects, John Dykstra, John Stears, Richard Edlund, Grant McCune, and Robert Blalack; music, John Williams.

Cast: Mark Hamill (Luke Skywalker), Harrison Ford (Han Solo), Carrie Fisher (Princess Leia Organa), Alec Guinness (Ben Obi-Wan Kenobi), Peter Cushing (Grand Moff Tarkin), Anthony Daniels (C-3PO), Kenny Baker (R2-D2), David Prowse (Lord Darth Vader), James Earl Jones (voice of Darth Vader), and Peter Mayhew (Chewbacca).

PLOT SUMMARY

The film opens with a prologue that provides background on the struggle taking place "a long time ago in a galaxy far, far away" of the Rebel Alliance against the evil Imperial forces led by Grand Moff Tarkin and his lieutenant, Lord Darth Vader.

The first scene begins with an Imperial spaceship chasing and firing at one of the Rebels' ships, which is carrying Princess Leia Organa, a Rebel Alliance Leader. The Princess, who has stolen the blueprints of the Empire's prime weapon, an enormous battle station called the Death Star, hurriedly transfers this data into the memory banks of a small android, R2-D2. The little android, ignoring the protests of its larger companion droid, C-3PO, makes its way to an escape pod which the two use to jettison from the Rebel ship just as the Imperial forces board and capture Princess Leia.

The two droids land on the desert planet, Tatooine, where they are captured by small scurrying creatures in monklike robes, the Jawas. The Jawas sell the droids to a local moisture farmer and his nephew, Luke Skywalker. As Skywalker is cleaning R2-D2, he discovers part of a holographic message that Leia has entered in the droid, a plea for help to an Obi-Wan Kenobi. During the night, R2-D2 runs away from the farm to deliver his message, and in the morning, Skywalker and C-3PO follow him. Just as they find R2-D2, the three are attacked by the fierce Tuskan

Raiders, also known as Sand People, who are about to kill Skywalker when the old hermit Ben Kenobi appears and frightens them away.

In the safety of his home, Obi-Wan (alias Ben) Kenobi hears Princess Leia's full message which explains the Rebels' dire situation. Kenobi tells Skywalker about the Jedi knights and "the Force," an energy field that can be harnessed for good or evil. He also tells Skywalker that Skywalker's father was a Jedi knight who was killed by Darth Vader. The dark side of the Force turned Vader from a Jedi knight into a servant of the evil Empire. Kenobi gives Skywalker a light saber that had belonged to Skywalker's father, and tries to convince the young man to help rescue the Princess. Skywalker reluctantly declines, saying that his uncle needs him to continue working on the farm.

When Skywalker, Kenobi, and the droids return to the farm, however, they discover that it has been burned by Imperial soldiers searching for the two droids (Vader has deduced that the stolen plans must have been jettisoned from the spaceship), and his aunt and uncle have been murdered. Skywalker then agrees to help Kenobi, and the four set off for a nearby spaceport to find a pilot who will transport them off Tatooine to the Rebels' home planet, Alderaan.

In a lively cantina filled with a great variety of life forms, they meet and hire mercenary pilot Han Solo and his copilot Chewbacca, a 7-foot-tall hairy being. Soon after taking off in Solo's spaceship, the *Millennium Falcon*, the group discovers that Alderaan has been destroyed by Grand Moff Tarkin—as an example of the power of the Empire. The *Falcon* then encounters a lone enemy fighter, which it chases toward a small planet. The planet is actually the Empire's battle station, the Death Star, and it uses a tractor beam to pull the *Falcon* into its hangar.

The group hides from the Imperial troopers who board the *Falcon*, and Skywalker and Solo ambush the soldiers and don their uniforms. Skywalker and Solo rescue the Princess while Kenobi disengages the Death Star's tractor beam so that the *Falcon* can escape. Solo, Skywalker, Leia, and Chewbacca become trapped in a garbage disposal unit while fleeing from Imperial soldiers, and are apparently doomed as the unit begins compacting. They are saved when R2-D2 shuts off the power.

Kenobi, meanwhile, confronts his former student, Darth Vader, and the two engage in a light-saber duel. When Kenobi sees that the Princess and the others are making their way back to the *Falcon*, he allows Vader to deal him a fatal blow, releasing his "force" into the universe.

The *Falcon* takes off and flies to Yavin, the Rebel Alliance's secret base, where the Rebel leaders analyze the Death Star's blueprints and discover that the only way to destroy the Death Star is to fire a laser bolt into a duct that leads directly into the center of the battle station, setting off a chain reaction. The mission must be accomplished using small fighter planes. Skywalker attempts to persuade Solo to join the Rebel mission, but Solo refuses, saying that he only wants the reward promised him for rescuing the Princess.

The Rebels and Imperial forces engage in a dog fight in space, with the Empire eventually destroying most of the Rebel fighters. Just as Darth Vader himself is about to shoot down Skywalker's fighter, Solo appears and sends Vader's ship spiraling away into space. Skywalker hears Kenobi's voice telling him to "use the Force" and he is able to fire the shot necessary to destroy the Death Star. In the final scene, Skywalker, Solo, and Chewbacca are honored in a ceremony at which they are decorated with medals by the Princess. They and the two droids turn to face the audience which cheers their valiant efforts against evil.

SUPERMAN

Warner Brothers (US/Great Britain), 1978, color, 142 min.

Credits: Producer, Alexander and Ilya Salkind; director, Richard Donner; story, Mario Puzo; screenplay, Mario Puzo, David Newman, Leslie Newman, and Robert Benton; cinematographer, Geoffrey Unsworth; music, John Williams.

Cast: Christopher Reeve (Superman), Marlon Brando (Jor-El), Margot Kidder (Lois Lane), Jackie Cooper (Perry White), Glenn Ford (Jonathan Kent), Gene Hackman (Lex Luthor), Valerie Perrine (Eve Teschmacher), Ned Beatty (Otis), and Susannah York (Lara).

PLOT SUMMARY

It is June 1938, in the decade of the Depression. *The Daily Planet* is one of our country's great metropolitan newspapers.

Far away, we see a giant red sun orbited by an eerie blue planet named Krypton. On the surface of the planet stands a futuristic dome in which three people are being tried for treason. The faces of the jury panel are magnified inside the dome as they pass down their ruling of guilty. Then, suddenly, the dome opens and a beam of light from above captures the three criminals and imprisons them in a space-bound two-dimensional plane.

The prosecutor of these criminals, Jor-El, knowing that Krypton is about to explode, then tries to convince its ruling council to evacuate the planet, but is accused of insurrection and told to remain silent. He goes home and plans with his wife to send their infant son to Earth. After putting some clear crystals into tubes, Jor-El places a special glowing green crystal into a star-shaped capsule with his son, Kal-el, and launches the capsule into space. As the rocket leaves, a series of earthquakes occur, ending in the spectacular explosion of the planet.

As the rocket travels towards Earth, we hear Jor-El's voice instructing his son about the contents of the spaceship. There are crystals containing the total accu-

Superman. Jonathan Kent and his wife are amazed at the strength of the young child they rescued from a crashed alien space capsule.
(Photo: Museum of Modern Art/Film Stills Archive. Courtesy of Warner Brothers.)

mulation of all literature and scientific facts known about 28 galaxies. The infant is told that his senses of sight and hearing will be exceptional and that he will have limitless speed. He is expressly forbidden to interfere with human history.

His capsule lands in a field as a middle-aged farmer and his wife drive along the road nearby. They pull the child and his spacecraft out of the crater. The woman wants to keep him, and her kind but sickly husband agrees when the boy lifts up their car after it breaks down returning to their Smallville home.

As a teenager, the boy (now known by his Earth name, Clark Kent) demonstrates that he can kick a football out of a stadium and outrun a train, but his adoptive father reprimands him for showing off. The sudden death of his adoptive father torments the youth because he could not save his father's life, despite his powers.

Some time later, the green crystal instructs Kent to move on. He tells his mother that he is going north. When he arrives at the North Pole, he throws a clear crystal into the snow, causing an explosion. Gigantic crystals grow out of the ice, forming the Fortress of Solitude. Over the years that follow, a holographic image of Jor-El takes his son on a journey through time and space. Jor-El explains that the red

Krypton sun is the source of the young man's strength as well as the cause of Krypton's destruction. After completing his education, Clark Kent emerges from the fortress with his superhero identity, that of Superman. He flies off.

Clark Kent then takes a job at *The Daily Planet* where he befriends reporter Lois Lane. Kent acts like a klutz until the day he secretly saves Lane's life by catching a purse-snatcher's bullet with his bare hand.

Meanwhile, we see villian Lex Luthor living in a swank estate hidden under the city's subway. The hideout contains state-of-the-art gadgets, an immense library, and a swimming pool.

In a crime-of-the-century scheme, Luthor has bought up all the desert land east of the San Andreas fault. He plans to divert a nuclear missile (about to be test fired) to strike the weakest part of the fault, thereby causing earthquakes that will push California into the ocean. This would create a new coastline owned by Luthor.

Lois Lane tries to take a helicopter from the roof of *The Daily Planet* building in order to meet Air Force One, but the helicopter gets entangled with a loose wire on the roof. The chopper ends up dangling over the edge of the roof, with Lane barely holding on. Kent, unable to find the traditional phone booth, changes into his Superman costume in a revolving door and flies up to catch her with one hand as she plunges to certain death. With his other hand he catches the falling chopper and he returns both to safety on the roof. This is his first public appearance as Superman.

Next, Superman apprehends a thief scaling a building, captures criminals escaping on a boat, and rescues a little girl's cat which is caught in a tree. When Air Force One is hit by lightning outside of Metropolis and starts to fall, Superman steers the jet to safety.

The editor of *The Daily Planet* orders his staff to get the real story on Superman. Now in love with Lane, Kent arranges to meet her at her apartment for dinner. While Lane waits for Kent on her balcony, Superman flies in, answers questions about himself, and then takes Lane flying.

Lane's article, entitled "I Spent the Night with Superman," is front page news for *The Daily Planet.* Luthor reads the article and announces that when Krypton exploded, fragments from the planet became meteors. One of these meteors crashed on Earth and is on display in a museum. He asserts that its substance is fatal to Superman.

Luthor and his accomplices redirect two missiles to be test fired. One of the missiles is now aimed at the San Andreas fault; the other is pointed at Hackensack, New Jersey.

While talking to his editor, Kent (and all the dogs in the neighborhood) hears an ear-splitting, high-pitched sound. Lex Luthor's voice tells Superman that in 5 minutes a capsule filled with deadly propane lithium gas will be detonated. Kent flies from the building, turning into Superman, and follows the voice to Luthor's hideout. There is a confrontation in which Luthor divulges his plan, and tricks Superman into opening a lead-lined box containing the stolen Kryptonite. Exposure to the Kryptonite weakens Superman and he falls into the swimming pool, but

Luthor's girlfriend rescues Superman after he promises to first save her mother, who lives in Hackensack, the second missile's target.

Lane is in the middle of the desert covering a land-scam story when her car runs out of gas. Meanwhile, Superman successfully intercepts the Hackensack missile and shoves it into outer space. Unfortunately, the other missile hits the San Andreas fault, which starts to cave in. Superman enters the fault and lifts up millions of tons of earth. Even though he has repaired the fault, earthquakes and landslides follow. Superman flies around saving lives. He stretches out over a damaged train track and lets the train roll over him; when a dam bursts, he prevents the water from flooding the land by throwing boulders into its path.

By the time Superman finds Lane, her car is buried in a landslide and she is dead. Superman decides to turn back time to prevent Lane's death, despite Jor-El's warning not to interfere with human history. Superman circles the Earth, reversing its rotation, and thereby causes time to move backwards. When he returns to Earth, Lane is sitting in her car before the landslide occurs. The movie ends as Superman turns Luthor and his associate over to the police.

TERMINATOR 2: JUDGMENT DAY

Carolco (US), 1991, color, 139 min.

Credits: Producer/director, James Cameron; screenplay, James Cameron and William Wisher; cinematographer, Adam Greenberg; special effects, Thomas Fisher; music, Brad Fiedel.

Cast: Arnold Schwarzenegger (Terminator), Linda Hamilton (Sarah Connor), Edward Furlong (John Connor), Robert Patrick (I-1000), and Earl Boen (Dr. Silberman).

PLOT SUMMARY

The movie opens with scenes from the year 2029 in which a war is being fought between the survivors of a 1997 nuclear holocaust and the sentient machine, Skynet. Skynet had initiated the 1997 World War in order to destroy the human race. Skynet sends an I-1000 robot back in time to kill the boy, John Connor, who will lead humanity in its struggle against the machines in 2029. The humans send back a robot, a reprogrammed Terminator, to protect the boy.

The story refers to the earlier film, *Terminator*, in which Skynet had tried unsuccessfully to have a Terminator kill John's mother, Sarah Connor, in 1984 before she gave birth to him. The earlier attempt was foiled when a human from the future aided Sarah in destroying the Terminator. He also fathered John before being killed fighting that Terminator.

John Connor is 10 years old as the film commences. Sarah has been committed to a mental institution for the criminally insane because she tried to blow up a computer facility. She had associated with mercenaries and revolutionaries since her 1984 experience with the Terminator from the future. Arnold Schwarzenegger, who played the Terminator in the earlier film, portrays a cyborg of the same model from the future. However, this time it is the protector of John Connor. The cyborg has living flesh over a metal skeleton that is very difficult to destroy.

John and a companion illegally extract $300 from a bank's automated machine and proceed to a video games arcade to spend their loot. Meanwhile, the Terminator has acquired clothes, a motorcycle, and weapons. It contacts John's foster parents but they do not know his whereabouts. The I-1000, which has assumed the appearance of a police officer it has killed, also visits John's foster home. It later follows John to the arcade where it chases him into a back corridor and has its first encounter with the Terminator. The powerful shotgun wielded by the Terminator demonstrates that the I-1000 will be difficult to destroy. The rifle pellets make temporary holes in the robot, but the creature promptly reforms itself. We later learn that the I-1000 consists of a kind of liquid metal that can assume the shape of anything with which it comes into contact.

John escapes from the I-1000 on his motorbike, but the robot requisitions a truck and follows the boy in one of the most spectacular chase scenes ever filmed. The boy is then rescued by the Terminator. They learn that his foster parents have been killed by the I-1000 and the boy orders the Terminator to assist him in freeing his mother from the insane asylum.

Meanwhile Sarah Connor has almost convinced her psychiatrist, Dr. Silberman, that she is a changed woman and thus can see her son. When the doctor decides that she is just telling him what he wants to hear, she attacks him in a rage. Later, the police show her a picture of the Terminator which was taken near the video arcade: the police believe it is a photograph of the same man who wiped out an entire police station in 1984 in order to kill Sarah Connor. Connor says nothing, but steals a paper clip with which she escapes from her room when both the Terminator and the I-1000 arrive.

Sarah, John, and the Terminator just manage to escape the I-1000. After treating a wound suffered by Sarah, the three drive south to a junk yard in which Sarah had earlier hidden weapons and supplies. As they unearth the weapons cache, Sarah observes that the Terminator is the perfect father for her son: it will always have time for John, will always be there to protect him, and if necessary will die for him. The cyborg is a better father than any of the men she has been with since John's birth.

As they prepare to cross the border into Mexico, Sarah again dreams of the terrible "Judgement Day" in 1997, when billions of people will die. There are spectacular special effects depicting a hydrogen bomb exploding over Los Angeles. When she awakens Sarah decides to prevent this calamity by killing the inventor of the Skynet computer chip.

Sarah wounds the inventor but is unable to bring herself to kill him. The Terminator and John then arrive and convince the inventor to help them destroy his research work which is based upon the computer chip taken from the remains of the Terminator killed by Sarah in 1984.

The scientist, Sarah, John, and the Terminator force their way into the research lab, from which they retrieve the chip and the mechanical arm of the 1984 Terminator. After a furious battle against dozens of police, the Terminator gets Sarah and John away from the research laboratory. The scientist is mortally wounded, and by a swat team all but destroys all of the research on the chip in a huge explosion.

The I-1000 follows them and, after a lengthy battle in which the Terminator is badly damaged, the I-1000 is destroyed by being blasted into a tank of molten metal: the liquid metal of its body dissolves into the molten metal. The 1984 Terminator's chip and arm are also thrown into the tank, and then the Terminator asks Sarah to lower it into the fiery cauldron in order to prevent humanity from using its chips to build a future Skynet computer. At the end of the film Sarah Connor muses that if a machine like the Terminator could learn to value human life, perhaps humans themselves will avoid a nuclear holocaust.

ZARDOZ

Twentieth-Century Fox (US), 1973, color, 105 min.

Credits: Producer, director and screenplay writer, John Boorman; music, David Munrow; photography, Geoffrey Unsworth.

Cast: Sean Connery (Zed), Charlotte Rampling (Consuella), John Alderton (Friend), Niall Buggy (Zardoz, Arthur Frayn), and Sara Kestelman (May).

PLOT SUMMARY

It is the year 2293. A giant levitated rock face lowers itself to the ground. A group of horsemen, called Exterminators, circle the head. One of them yells, "Praise be to Zardoz!" They are given guns and told to kill the brutals.

One of the Exterminators, Zed, smuggles himself aboard the giant stone head of Zardoz. Zed investigates the interior of the head, now airborne, and finds humans encased in plastic wraps. He then sees a man walking near the open mouth of the head and shoots the man. As the man, who turns out to be Zardoz, falls from the head, he warns Zed that without Zardoz's godly powers, Zed is nothing.

The head then descends into a rural area. Zed emerges from the head, goes into a nearby house and discovers a shrine to Zardoz. The house belongs to Arthur Frayn the man who Zed has just shot. Zed picks up a seemingly magic communication crystal and leaves.

When Zed sees a woman on a horse, his warrior/exterminator instincts take over, but a force prevents him from harming her. She tells him that he is in a vortex where people live forever. He tells her that he is an Exterminator for Zardoz.

The people ("Eternals") from the strange society inside the vortex examine Zed. His memories are reflected on the walls. We learn that Arthur Frayn died three days before: reconstruction of Frayn has already begun. The group learns that Zed's mother and father were chosen to breed. Zed looks forward to the moment of executing the brutals when he is "one with Zardoz." The group decides to have one of the Eternals, May, study him for three weeks.

In this society people do not age, unless they are punished for committing a criminal act. Zed sees a man on trial for psychic violence. When the man is found guilty, he is "aged" an appropriate number of years. People can never die because the Eternal Tabernacle continues to rebuild them. Some people are aged to the point of senility. Others, tired of living in an eternal state of youth, become Apathetics. Eternals also have no need for sleep. Their waking and unconscious lives are one.

Time passes: at a dinner, the Eternals vote to destroy Zed in seven days. They then meditate, but one of them, Friend, does not want to become "one" with them. He is punished by being aged.

Zed finally explains to May that Zardoz gave the Exterminators guns to hunt and kill. The Eternals later decided that they needed the brutals to produce wheat in order to feed the nonworking Apathetics and thus Zardoz ordered the Exterminators to supervise the cultivation of wheat by the brutals. Zardoz came down each season to collect the harvest.

Zed, who had been taught to read by the man Zardoz, discovered that the name "Zardoz" comes from The Wizard Of Oz. Zardoz, like the Wizard, was not a god but a man behind a "mask." Zed then hid aboard the stone head to learn more about Zardoz.

Zed is told that he has been chosen to be the one to liberate the Eternals from their world of perpetual life. He is told that before the vortex was created the world was dying. The Eternals took all that was good and built an oasis, surrounded by a force field. They hardened their hearts to what was going on in the outside world because they knew that they were the custodians for an unknown future. Zed realizes that in order to restore the act of death to the Eternals, he must get inside the Tabernacle, which controls the vortex.

Scientists invented the Tabernacle and erased the memory of how it was made. Zed discovers that the Tabernacle is a crystal, which he enters and destroys from within. Death is then restored to the vortex as the head of Zardoz crashes to the ground. The Exterminators enter the vortex and slaughter the "immortals," who welcome death. Zed and one of the former Immortals, Consuella, escape to a secluded cave where they have a child, grow old together, and die as the film ends.

Index